Tactical Lock Picking Tool & Technique Guide

WARNING

The author and publisher are not liable for any injuries, damages, or losses resulting from the use or misuse of the techniques described in this book.

Readers are strongly encouraged to seek professional training, and to exercise caution when using any sharp or dangerous tools.

Access only those locks for which you have been given explicit permission by the rightful owner, or when you have otherwise attained permission both morally and legally.

To purchase many of the tools seen in this guide, and to sign up for one of our in-person courses, check out the QR code above, or visit UncensoredTactical.com (or the shortcut UTAC.io)

Email: Pat@UTAC.io
Website: UncensoredTactical.com & UTAC.io
Socials: @UncensoredTactical

ISBN: 978-1-7348254-8-0 (Paperback)
ISBN: 979-8-9919801-0-4 (Spiral Bound)

Ordering Information:

Special pricing and permissions are possible, with our written approval, for distribution/reselling, and for educational orders. Please contact the email above for inquiries.

UTAC.io

TARGET ASSESSMENT STEPS:

1. Do you have the **Moral Right** to make entry?

2. What is the **Speed** at which entry is required?

3. What **Attack Vectors** are known to you at this time?

4. **Manual Check** locked/unlocked status. (Then take a breath, then try again harder)

5. **Tool Availability**? What tools are on you, near you, can be retrieved by you or brought to you, or can be makeshifted within your environment?

6. **Now or Later**? Should you start your hands on entry attempt now, or do more recon and then decide which vector you will attack through?

7. **Other Resources**

- Internet **Research**, locksmith, phone-a-friend.
- Take a big **Reset breath**.
- **Retrace** all of your steps so far.

This field manual is dedicated to those who wish to learn.

Whether you intend to learn this skill as a hobby,
at a top-tier operational level,
or anywhere in between...

*We are here to **teach**, and we **welcome** you to our craft.*

TABLE OF CONTENTS

1 THREE LINES OF GEAR

5 RAKING & SINGLE PIN PICKING (SPP)

15 PADLOCK BYPASSES

27 DOOR BYPASSES

39 LISHI 2-IN-1 PICK/DECODERS

43 COMB TOOLS

47 COMBO LOCKS - PHYSICAL DECODING

49 COMBO LOCKS - SOCIAL DE-ENGINEERING

51 BRUTE FORCE CODE ATTACKS

69 KWIKSET SMARTKEY LOCKS

77 WARDED LOCKS

79 WAFER LOCKS

83 KEYED ALIKE

85 VEHICLE ENTRIES

87 TARGET ASSESSMENT

93 LOCK IDENTIFICATION

99 MAKESHIFT TOOLS

103 THANK YOU

BYHALIA, Miss. (March 2024)

THREE LINES OF GEAR

These are not rules! None of our Tactical Lock Picking curriculum is ever framed as "rules you must follow," but simply as information for you to consider while building your own unique skillset.

1st Line Gear (1LG) the things in your pocket, in your wallet, and on your belt (EDC).
2nd Line Gear (2LG) a small go-bag with larger sizes and selections of entry tools.
3rd Line Gear (3LG) the physical tools that otherwise don't fit in the other two lines of gear; often considered your vehicle storage or closet storage.

This "Three Lines of Gear" concept is just *a way* to think about how to *deploy* your tools into the field for operations or for emergencies. You can mix and match, you can have duplicates, you can have something that doesn't fit our definitions, and we encourge that creativity and that customization.

Don't get hung up on forcing this system, just use it as a foundation to get your feet wet while you are building and adjusting your skill set to work for *you*.

1LG:

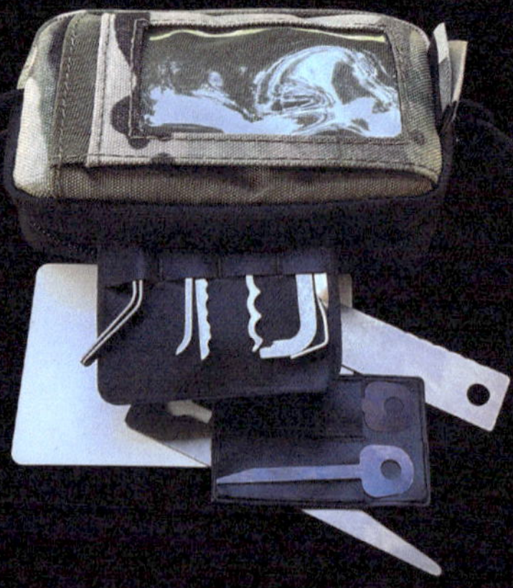

Recommended Foundations:

- At least two different rakes.
- At least two different tension wrenches.
- At least two different hook/lifter picks.
- Rigid tool for pull door latch slipping.
- Flexible tool for push door latch slipping.

Additional Considerations:

- EZ Decoders for combo decoding.
- Wafer jiggler tools.
- Warded pick tools.
- Comb tools.

2LG:

Recommended Foundations:

- Full sized door jims (pull door).
- Full sized door shims (push door).
- Expanded set of wafer jigglers.
- Expanded set of warded picks.
- Bypass drivers.
- Extra EZ Decoders.
- Additional full sized picks/rakes.
- Additional tension wrenches.
- Selection of Lishi Decoders.
- Plug spinner.
- Brute Force Notebook (UTAC.io).

Additional Considerations:

- Key generation tools.
- Administrative tools (multi-tool, flashlights, wire cutters, wedges and pry bars, etc.).
- DoorJamm - for keeping and sharing access.).

3LG:

Recommended Foundations:

- Destructive entry tools (sledge, pry bars, glass breakers, etc.).
- Vehicle reach tools.

Additional Considerations:

- Administrative tools (multi-tool, flashlights, wire cutters, etc.).
- Kwikset Smartkey - Tryout Set (UTAC.io).
- Air wedge tools.
- Key generation tool case.
    - Smartkey camera.
    - Key blanks.
    - Hand files.
    - Key gauge card.
    - Etc.

- Remember that these are guidelines, not rules, and do not need to be memorized.
- Gear stacking is individual to each operator (or each team). The differences in organization and specific terminology used to explain this setup is not important, these are merely concepts to assist you.
- You should move items fluidly between your lines of gear as necessary. How you organize your gear and how you choose to move with your gear is a valubale part of our overall skill set teachings.
- Your gear system should be alive and open to evolving. The end goal is always to support problem solving in the field. Changes in skill, budget, technology, and lessons learned, should all help to shape your gear loadout evolution.
- Changing your gear loadout too quickly, too often, or for poor reasons can hamper field effectiveness; leaving you to spend time figuring out "did I actually pack that tool or not?" and searching through your gear. With abrupt changes, you lose the ability to make rapid informed decisions regarding tool & technique application.
- And keep maintainence in mind. How are battery powered items stored and when are they charged? Do you have charging cables? Will you need extra battery banks for them? Does any of your gear require cleaning or polishing? Is all of your gear waterproof, and if not, how will you package that into your kit with that in mind? Etc.

Non-Destructive Entry (NDE) in the field, of any type, is an entirely different practice and an entirely different skill set than hobby picking or desk picking.

At a desk, or in a classroom, or on your couch, the tools and techniques you choose exist within their specific environment. Those desk/classroom/couch decisions have a zero percent likelihood of having a negative consequence. You can take all the time and use all the resources that you choose in those relaxed settings.

In a real world emergency or operation, the nature of "real world" means that your resources will all be limited in at least some way, and all of your choices will have consequences. Tactial Lock Picking (TLP) is a numbers game, a thinking game, and a problem solving game.

It is still a successful TLP entry in the field if you decide to use something other than non-destructive means. Even an "abort mission" decision can be a win as a TLP specialist if the decision is reasonable for the operation. A "win" for UTAC students is always earned if the decision on scene is *effective*, *efficient*, and *reasonable*, regardless of whether it was destructive, non-destructive, or other. The goal is better operational outcomes, not tunnle vision for a particular method of application.

ZERO LINE GEAR:

This is **information** that you have on hand; memorized, or printed reference cards, or other info resources.

Some examples: information that you can access like websites, or search engine results, or phoning-a-friend for info.

.5 LINE GEAR:

Covert tools, or other tools similarly unique in small size and storage.

Some examples: a tiny Bogota® Nano set that is sewn behind a brand name tag on your pants, or safety pinned into a piece of clothing.

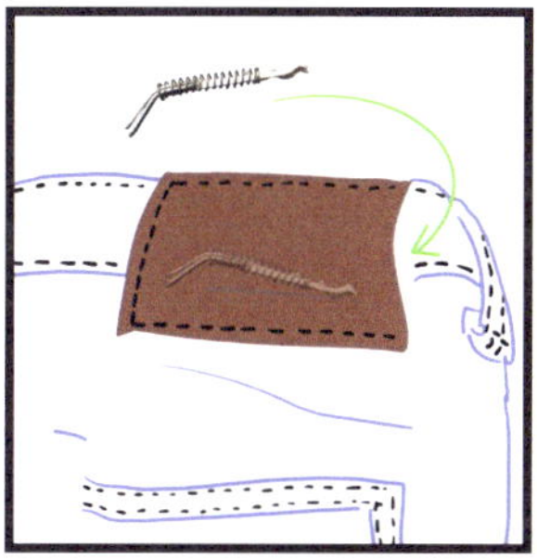

1.5 LINE GEAR:

SMALL UTILITY CARRY

Too large for your EDC, but smaller than a dedicated go-bag. Things like fanny packs; or a clipboard with a deep storage compartment. And if you own body armor, entry tools on your plate carrier wouldn't be EDC or a go-bag but an in-between.

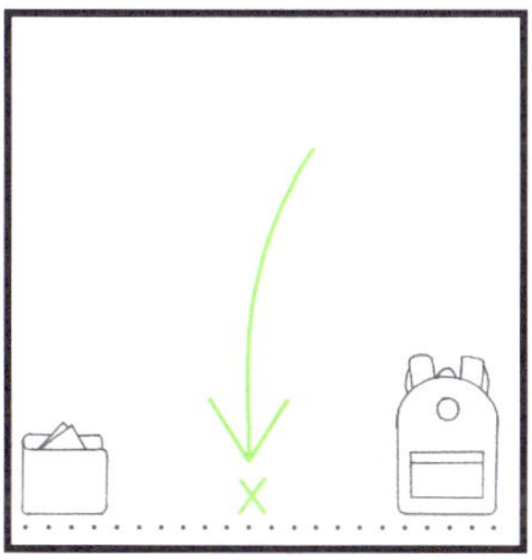

4TH LINE GEAR:

ENVIRONMENT

Tools from your environment that you can acquire. Wire coat hangers and chain-link fence wire (for reach tools); laminated paper or restaurant menus (latch slipping); aluminum cans (padlock shims); tools from maintenance, from other responders, or even from a nearby hardware store.

RAKING

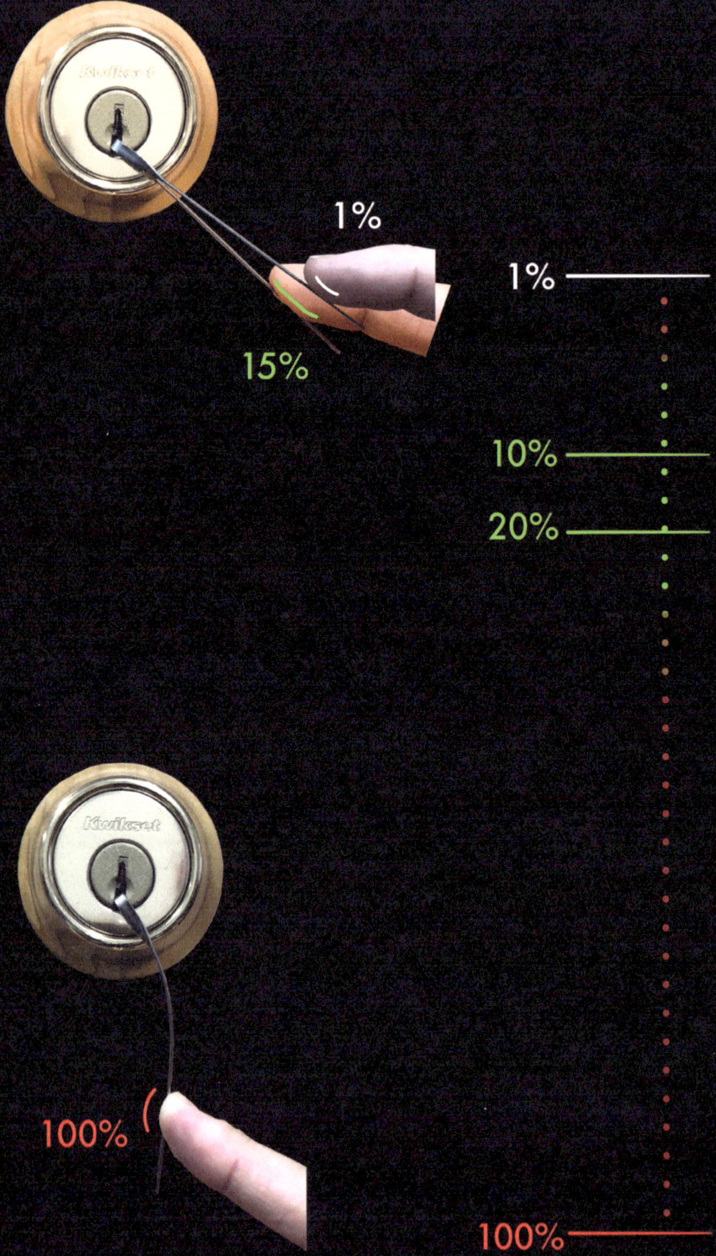

Apply rotational tension force with your tension wrench, while at the same time using your raking tool (or picking tool) to go in/out and up/down.

Depending on the lock, you will start to identify which direction to apply rotational tension to.

For your tension amount, If 1% pressure is barely touching the tension wrench, and 100% is bending the tool to a permanent bend, aim for around 10-20% pressure.

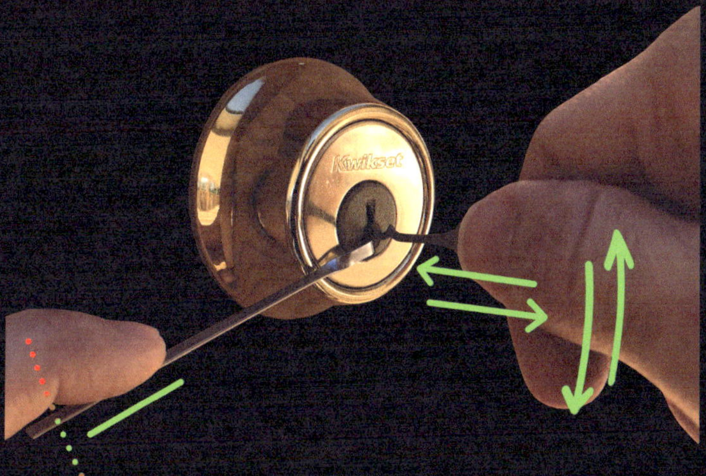

Since this raking method is quick to learn and can be applied very early in your skill journey, we focus on this method first before teaching Single Pin Picking (SPP). This raking method works on many common locks and low security locks, however, it is not as effective on higher security locks (like SPP is).

Our goal is to make people operationally ready as fast as possible with the most effective and efficient tools and techniques. Those tool/technique pairs should be in alignment with the most likely locked obstacles that you expect to encounter in the wild.

SINGLE PIN PICKING (SPP)

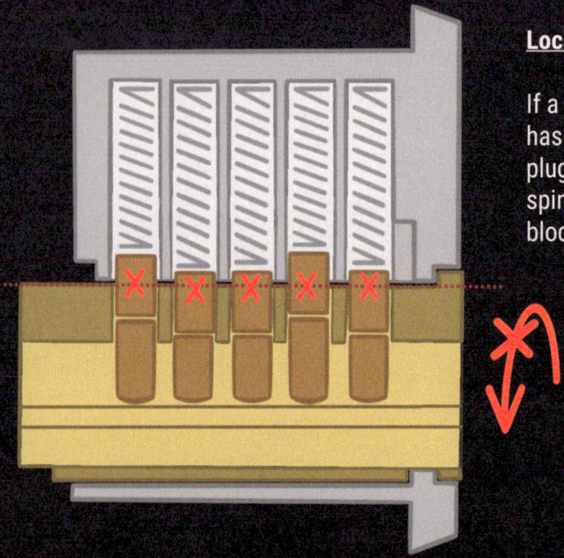

Lock at Rest:

If a pin-tumbler lock at rest has tension applied to the plug (the keyway) it will not spin. The driver pins will be blocking the shear line.

Correct Key Inserted:

With the correct key, the driver pins (the top pins, touching the springs) will be raised above the shear line, and the key pins will rest flush due to the cuts of the key. This removes any pin obstructions from the shear line, allowing the plug to spin, and the lock to open.

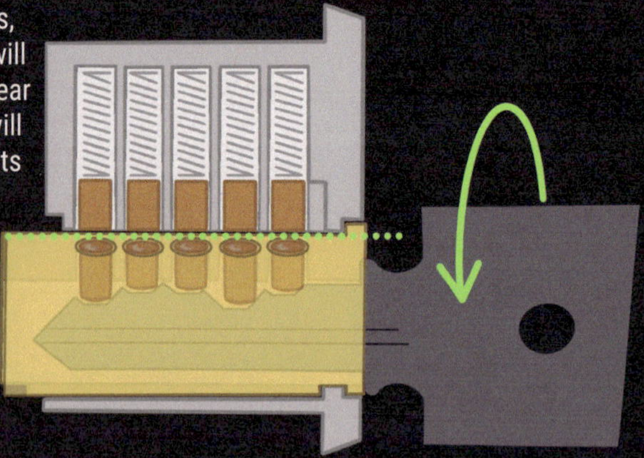

***Don't Get Discouraged!**
SPP takes a long time to learn and to become proficient at!

Find the First Binding Pin:

Apply a medium amount of tension in what you believe is the correct direction, [you'll learn more about this as you go]. Try to gently touch each of the pins to see if one of them is the most stuck/binding/sticky.

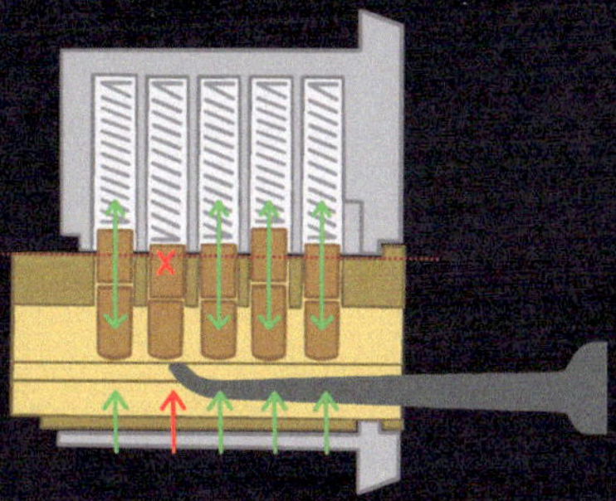

Lift the First Binding Pin:

Once you've identified which pin you think is binding (it may be different on each lock and it may be two pins that seem to bind), gently but firmly lift that pin stack until you hear/ feel a distinct *click*.

This specific pin stack, on a low security lock, probably doesn't need to be set again.

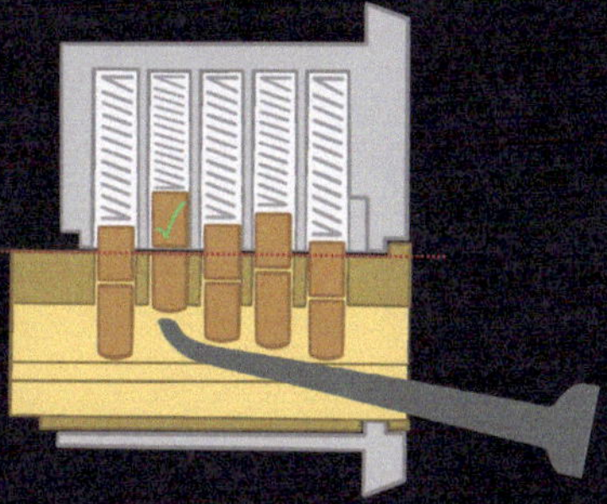

Find the Next Binding Pin:

The first pin stack that you've set should have the driver pin (top pin) resting above the shear line, and the key pin should fall to resting. Then locate the next binding pin stack (the other remaining un-set stacks should still be springy).

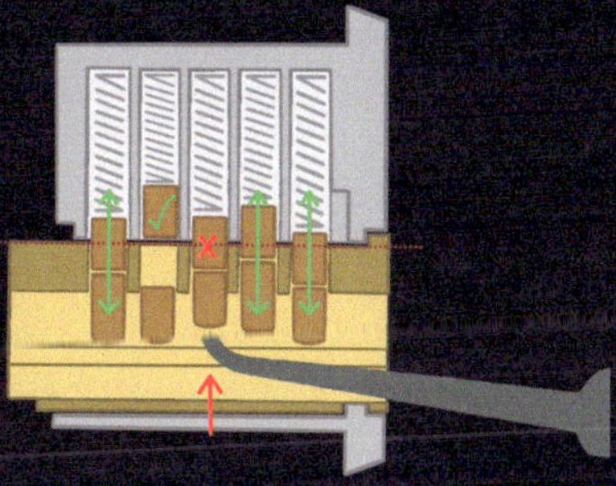

Continue & Troubleshoot:

Continue the process until all of the driver pins are above the shear line and the plug (keyway) spins open.

This is both a science and an artform that takes a long time to develop. Please remember, **do not get frustrated** or overwhelmed!!!

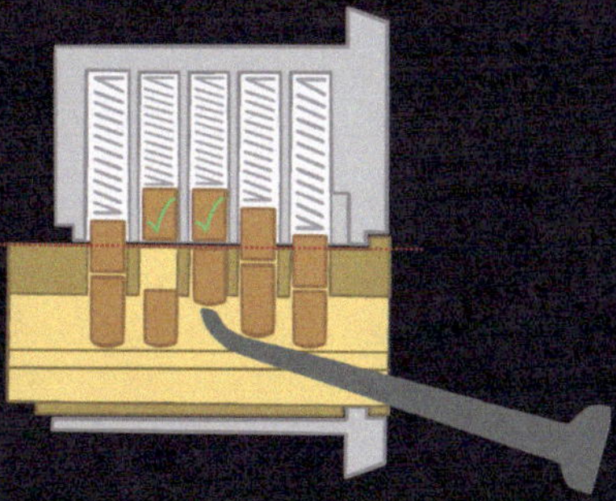

Note: This is a very general introduction to the concept of SPP, and should not be considered a complete learning block.

TENSION TIPS

Applying proper tension is not just about "which tool is best." Applying proper tension is about problem solving, and understanding expected and unexpected variables and how to adjust your approach and how to find *possible* solutions.

Troubleshooting:

- 9 out of 10 times, the reason raking doesn't work is that you're using too much tension.
- 1 out of 10 times, you should try more tension than usual (after you've exhausted attempts with less tension).
- It might be open and you might not know it. Occasionally, mash that tension wrench down.
- Tension is almost always more important than the raking speed and the raking motion.
- Do you have a different tension wrench that might provide a better solution?
- Sometimes keyways are easier to pick when tension is provided in the opposite direction. On occasion, a plug spinner might be able to help you get the keyway (the plug) spinning in the wrong direction first, and then flipping it to the right direction with a plug spinner.
- Is the lock actually "pickable?" Is it too high of a security lock for a basic raking attack? Is it *actually* a pin tumbler?[see Smartkey Chapter]
- Have you looked around yet to find a better entry attack vector than picking through the keyway, such as a latch slip or an open door or window or a realtor key-box nearby?
- Remember to practice placing tension wrenches into your keyway, sticking out of the keyway towards different directions (N,S,W,E or 12, 6, 9, 3 o'clock).
- Also, remember to practice using different hands and different fingers/thumbs for tension. The more you practice unique positions and unique solutions in training, the less you will have to experiment in the field.

For your custom skills toolbox we recommend the following:

If you don't yet know why your raking technique is working or not, and if you aren't familiar with how to work this tool/technique combo to its exhaustion through all possible variables, then you probably shouldn't start worrying about advanced tools/techniques yet. We're not saying *not* to buy expensive tools or to try difficult techniques, we're just recommending a training progression to build you into the most effective Tactical Lock Picking practitioner and also to save you time, money, resources, and frustration with an established training path.

TENSION TOOLS

RAKE TIPS

- Most beginners make the mistake of trying to *force* their rake towards the top of the keyway. This is rarely effective. Instead, consider "letting the rake and the pins talk to one another in a conversation."
- The movement of your rake is way less important than the application of the proper amount of tension, at the right time, with the right type of tension tool.
- The selection of "which rake shape/profile should I use?" is often secondary or tertiary for why a raking technique isn't working. Sometimes yes, but more than likely the problem is the technique of the practitioner. When trouble-shooting, the *height & thickness* of your rake tool is often more important than the detailed shape of your rake of choice. (Most advanced entry specialists have only two or three rake tools that end up opening 99% of all of their rake-able keyways)
- Some rakes will move around more smoothly inside of the keyway, and some rakes will feel rough or sticky or gritty. *This* is a factor that you should pay attention to, not just "which shape of rake should I use?" Learn to *feel* the personality of each tool and each lock you encounter.
- The handle and the grip of your rake tool is also probably more of a help to you than the actual shape of the end of the rake. There are full sized handles of different materials, pocket carry handles, and tiny covert handles (photo opposite: top row).

An overwhelming majority of high-level entry specialists from all walks (locksport, locksmithing, emergency first-responders, security specialists etc.) will carry one or two rake tools that they favor, and they will have dozens of rake profiles stored somewhere else collecting dust.

Raking tools are important, but tension is king.

Raking has many benefits where raking is often easier to learn, faster to get an opening (if it works), and usually doesn't require an exactly perfectly fitting rake tool - usually most types of rake profiles (shapes) will do.

Our recommendation for adding raking to your skills toolbox is to find a few rakes (and a few tension wrenches) that seem to be working for you. Use those as much as possible on as many locks as possible. Try to go as far as you can with your raking skill to find out what locks you *can* open with raking, but also what locks you *can't* open with raking. And stick with it for a while to find out how much you can stretch your raking skill.

SPP will certainly allow you to open more difficult locks of higher security, but raking will serve as a valuable foundation for you during emercency and operational entries.

RAKE TOOLS

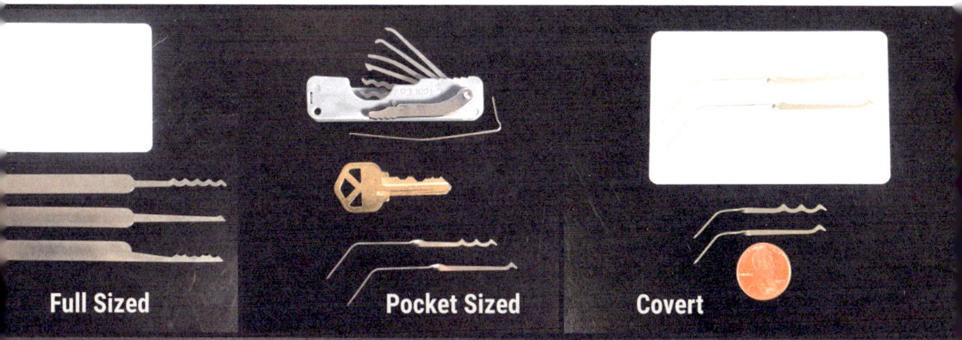
Full Sized
Pocket Sized
Covert

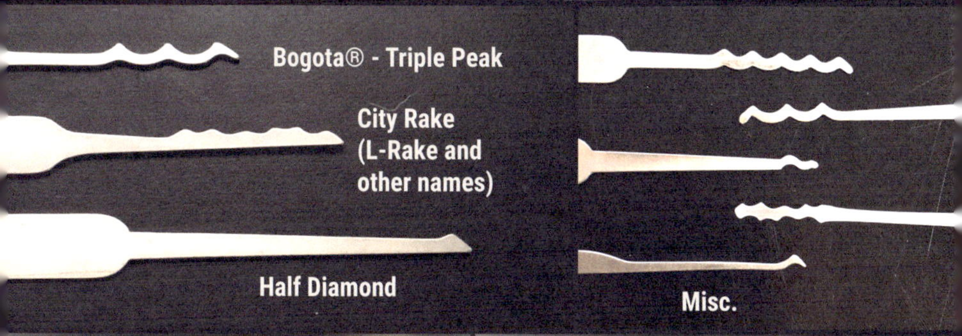
Bogota® - Triple Peak
City Rake
(L-Rake and
other names)
Half Diamond
Misc.

Thick
Thin
Thick
Thin

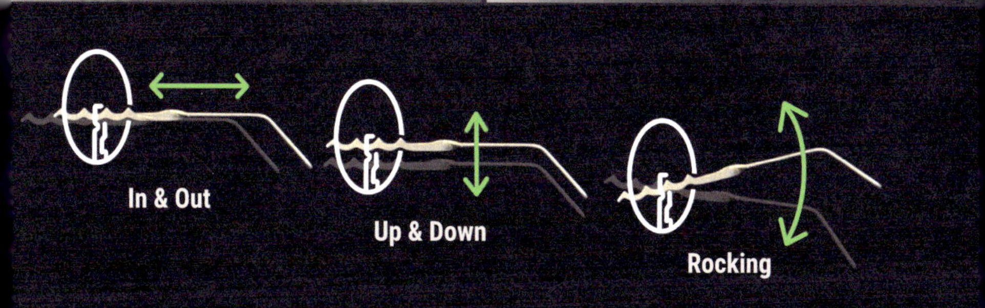
In & Out
Up & Down
Rocking

SPP TIPS

Single Pin Picking helps with higher security keyways. While SPP is much more scientific it is still an artform. SPP is more repeatable and reliable than the raking method. However, SPP takes a long time to learn and longer to master. It takes a lot of fine motor skills and an intimate knowledge of what's happening inside the lock. The tools required for SPP are much more specific than someone choosing a random rake for a raking attack.

Even experienced pickers have to troubleshoot picking a lock that is perfectly mounted to their work desk, with great lighting, controlled climate, and from a comfy rolling chair, etc. If simple and reliable tools/techniques end up requiring troubleshooting in a controlled environment, then imagine how troubleshooting might be more complicated in a stressful emergency or operation in the real world.

Just because you can pick a specific lock with a specific tool & technique at your desk, does not mean that you can repeat it with the same level of effectiveness during a field entry. And, just because you picked a specific lock open with a specific tool *once* does not mean that you can do it again, *especially* in field conditions.

Single Pin Picking is almost certainly worth adding to your skills toolbox, and when you do, here are some tips to consider:

- Single Pin Picks will break more often than rakes with regular use, pack your loadout accordingly.
- Top of Keyway (TOK) tension will suit SPP attacks better than bottom of keyway tension often, but not always.
- Sometimes SPPing in one direction of tension might be a struggle, consider picking to the other direction of tension and then adding something like a plug spinner into your toolbox.
- It's not just the shape of your hook pick, or lifter pick, that matters. It's also the thickness of your pick.
- Remember that the real masters of this craft recognize that *tension* is still the name of the game. Your tension hand deserves most of the credit even though the picking hand gets most of the glory.
- Don't get frustrated if you feel yourself not progressing fast enough. This takes time! And, social media influencers often lie (even if by omission) about their skill level, and their number of camera takes, and with how many failures they experience.

Photo Opposite: You will start to favor not only the profile of picks that you like, but also the design of the handles.

SPP PICKS & RAKES
COMMON FULL-HANDLE PROFILES

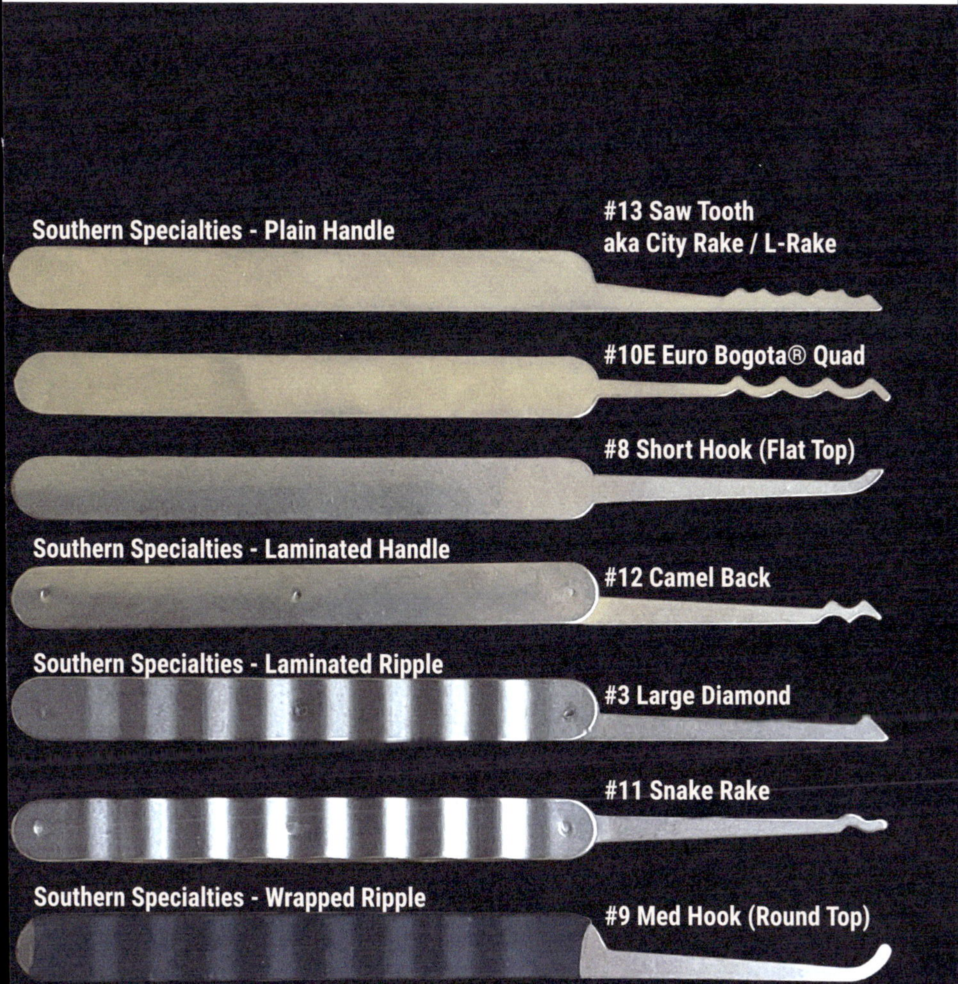

PADLOCK BYPASSES

One of UTAC's Tactical Lock Picking principles is understanding the **Cat and Mouse Game**. Once a feature for a particular lock is exploited enough, manufacturers may update the design of their locks. This attempt to protect against the exploit might be 100% effective, might make it *slightly* more difficult to exploit, or it can sometimes create a whole new separate exploit in the process.

This chapter has a selection of common padlocks that you will *definitely* encounter (in the U.S.), along with some tools & techniques that *might* open them. Remember that the value of our TLP curriculum is not just knowing how to use a tool in an *administrative* setting. The value is in knowing that there is a percentage chance that your tool/technique combo might *not* work; and the expert TLP operator knows how to make decisions and how to troubleshoot problems in the field when things *don't* work as planned and when resources are limited.

Padlock bypass tools are usually (not always) larger, bulkier, heavier, and more delicate than most of your standard pocket-sized pick sets. And while economics don't necessarily apply at your lock picking desk, the economics of budget management, gear storage capacity, and especially time management are *always* present in the field - where our TLP curriculum was born.

When approaching locks in the field, remember that bypasses are usually more efficient and more effective than picking (...not always) but bypass tools usually live in your larger 2nd Line Gear bag. One of the biggest and most common questions you'll run into as you develop your own case studies is the question of "Should I try to pick this lock now with the small limited tools I have in my pocket, or should I go get my 2nd Line Gear bag and come back with better tools before I start my entry?"

We are excited to hear about your trials and your successes and even more excited to hear about your failures and how you did or how you plan to overcome them. Please send your case studies (gathered and shared with proper permission) to my email: Pat@utac.io. Please help the UTAC team stay sharp by sharing with us and letting us know if our curriculum has helped you during your operations and emergencies.

There are many tools and many padlock bypass techniques that exist for you to consider adding into your skillset toolbox. Let this short chapter serve as a basic foundation of common examples for you to build upon. And remember that these are things that *might* work. When a tool/technique doesn't work on a lock, your job is to figure out if it's a tool issue, an operator issue, an environmental issue, or any other type of known or uknown variable. This is where you will sharpen your craft, learning how to solve problems.

MASTER 175

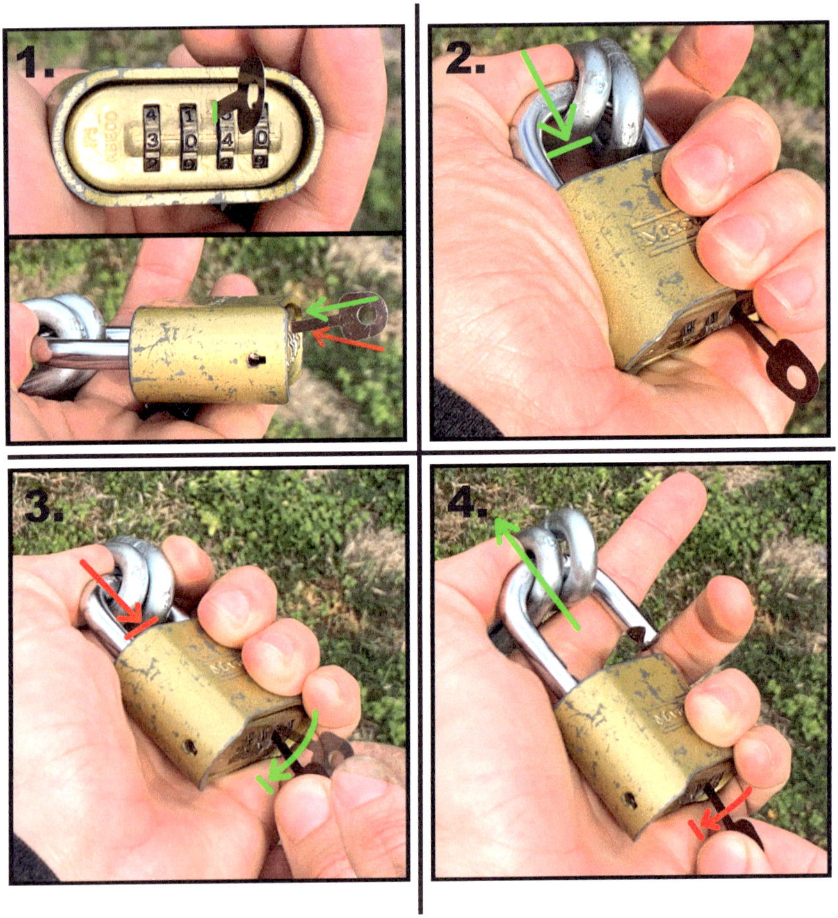

This is one of the first padlock bypasses we teach in our courses, and one of the most well known padlock bypasses even amongst new practitioners. As another note on nomenclature, there are common "types" of bypasses, sometimes so called because of the tools that perform the bypass and sometimes called such based upon the function within the lock. This bypass pretty much lives in its own camp of "bypass."

1. **Insert** your "EZ Decoder" vertically into the third wheel to the right (ensure the numbers on the wheels are not upside down), towards the top left corner of the wheel. The insertion depth only needs to be about an inch or so. Position the insertion angle downwards so that the tip is going in towards the center of the padlock (not up towards the outer shell of the padlock).
2. **Press the shackle in** *and hold*.
3. **While holding** the shackle in, **rock the tip of the tool up** (by pushing the handle of your tool down) *and hold* that upward pressure with the tip of your tool; rocking an internal locking plate into the up and open position.
4. **While holding** the tip of the tool up, **release the shackle**.

Lock identification is a big deal. This will be way easier on some Master 175's and similar locks, and may be much harder (or impossible) on other locks of the same or similar makes/models. This particular tool/technique combo works with a 90% likelihood or higher on older Master Lock 175's. The manufacturer has since tightened up the tolerances on new 175's and has also come out with a 178 (black coating) that adds a slight shielding within the lock, making it harder to slide your tool into place in the newer 178 models.

There is also a concept called "clones" in the lock world. Basically, a lock design is sent overseas and mass produced with the exact same, or almost the same internal and external design, just produced with a different name on the outside of the lock. This bypass technique *might* or *might not* work in clone locks.

This probably will not work on other types of locks that have different internal operating mechanisms. As you learn, you will naturally start to identify how to navigate these differences between similar makes and models. You will also learn that even the same makes and models of lock will have veried tool/tecnique effectiveness based upon their date of manufacturing (just like a Ford F150 from 1999 operates differently than a Ford F150 in 2019).

We place EZ Decoders into a classification of tools that we call "disposable." Not because they are poor tools but because they become damaged easily, and because they are important enough to store multiples within your kit, and because they are inexpensive. Buy multiples, carry multiples, don't feel bad when they break.

QUICK STICK BYPASS

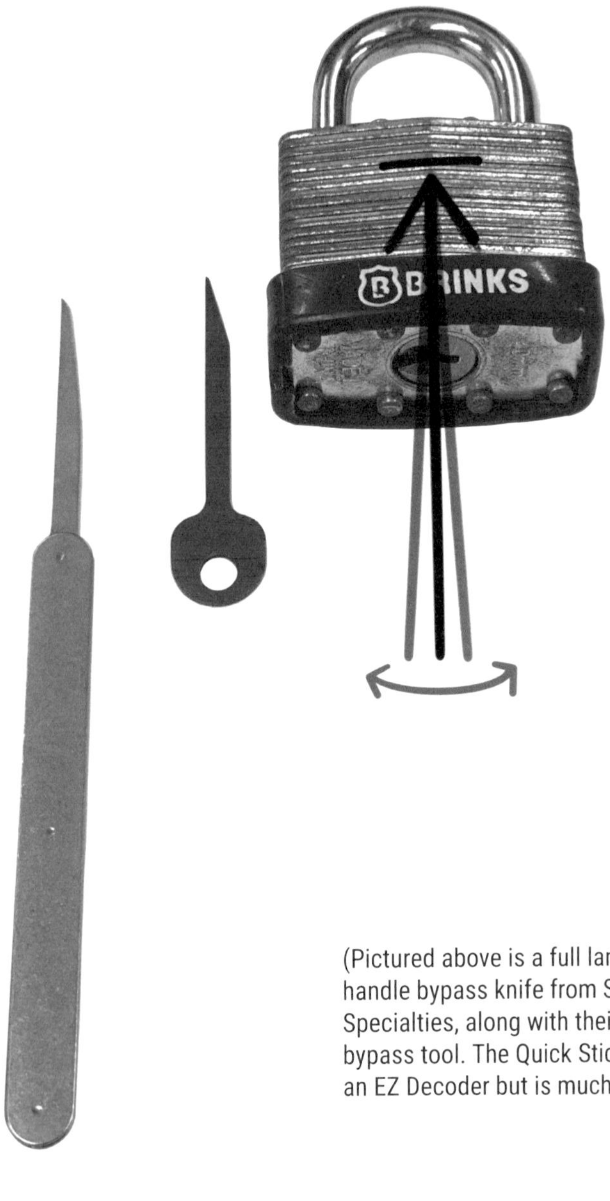

(Pictured above is a full laminated handle bypass knife from Southern Specialties, along with their Quick Stick bypass tool. The Quick Stick looks like an EZ Decoder but is much thicker.)

This method is possible on *multiple* makes and models of pin-tumbler padlocks, but not on *most* types of pin-tumbler padlocks. Most padlocks are manufactured with what is called internal "shielding." Shielding prevents tools from reaching through the back of the keyway and into the internals of the padlock. This Quick Stick bypass, also known as a "heel toe" bypass, is most likely to work on older and lower quality padlocks.

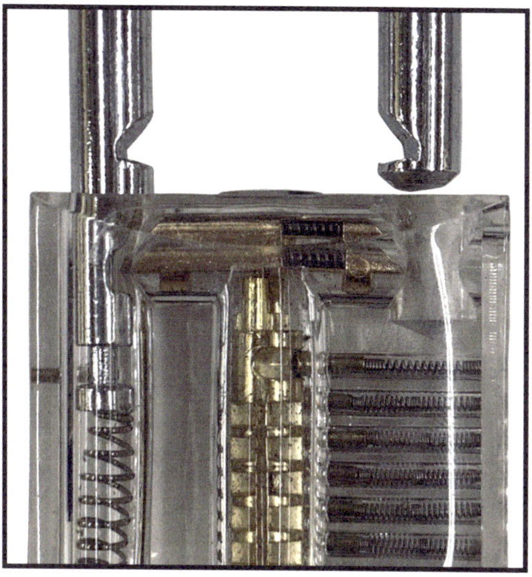

In the field, you might not know which direction you should be leveraging the quick stick towards. Our field guideance is "try both ways." You might not immediatly know if this is even possible on the current lock that you're dealing with, our guidance "Try all of the most likely and most reasonable tool and technique attempts in the most effctive and efficient order that you think will work for you." There are many padlocks that require *one* locking latch to release, and also many padlocks that require *two* opposite locking latches to be released.

When you have the right tool and the right lock, this technique usually works withen seconds. But, in our estimate, this works on less than 10% of the padlocks that we find in the wild. Because of this, our recommendation is to start with a raking technique or any other *known* bypass techniques first; then, if you struggle, consider trying a quick stick attack, along with other less-likely or other unique methods next. And remember, our field recommendations are just that, *recommendations*.

A small Quick Stick (shown in photo opposite) is able to be taped to the back of a credit card for very easy 1st Line Gear storage, or slid onto a keychain through the eyehole of the tool. A larger handled tool (shown opposite) helps to get the leverage that certain stubborn locks require with this type of attack, but due to their larger handled size they are less likely to be carried consistently in your pocket every day. The larger handled tool definitely deserves a spot within your 2nd Line Gear kit, but it's unlikely to be the most used tool in that bag.

- With your tool fully inserted to the back of the lock, keep forward pressure on the tool while rocking the tip of the tool high and low in the keyway. That's usually all that is required for this tool/technique combo.

BYPASS DRIVER

For locks without security wafers, the bypass driver goes past the keyway and wedges between the core's cam, and the padlock's cam. Then the bypass driver rotates, turning the cam inside the padlock, releasing the shackle.

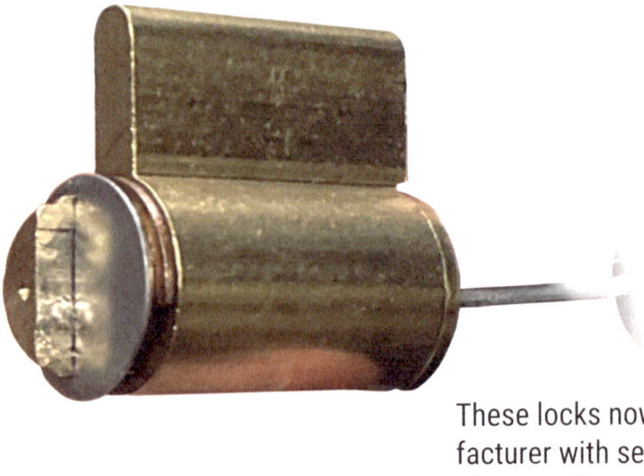

These locks now ship from the manufacturer with security wafers installed. These wafers make a bypass driver attack difficult, requiring extra tools and extra steps.

Simple, but not always easy. The first requirement is that you have the right tool on hand for the right make/model of lock. You also need to know that these American brand locks have been updated, and now shipped with security wafers installed (shown opposite: bottom). While these wafers will halt a bypass driver technique, there are tools that can be used to punch through wafers, but wafer breaking is not 100% effective either.

Additionally, once your tool is seated past the back of the keyway, sometimes it's hard to get leverage against the spring pressure of the padlock's cam - this often bends the tool, sometimes minimally and sometimes to a breaking point (below).

Bypass Driver Technique for American 1100 and 700 locks:

- Insert your bypass driver tool with the "flag" down towards the bottom of the keyway, all the way to the back of the keyway.
- Continue pushing in and wiggle/twist it a little until the tool "seats" itself a little farther to the back. The tool should now feel snug.
- You can check for a little springy-ness first to ensure you're positioned correctly, then twist the bypass driver at least 90-degrees to the right.

In the field, you can expect that about 20% of the time this technique works perfectly. 80% of the time it's a struggle or failure. These bypass drivers also fall into the category of **disposables**. These tools bend and break easily. Sometimes these tools will even show up directly from the manufacturer out of spec and won't seat correctly; we've even had brand new bypass drivers not even fit into a keyway on a known-correct lock.

There are multiple troubleshooting steps that you may need to take on bypass drivers attempts.Our field-guidance is that these are very useful tools *when they work*, and you should understand their limitations. Purchase multiple bypass drivers at a time and carry multiple drivers in the field. Be prepard to use pliers to re-straighten the working end of the tool if you find it has bent during use. These may also be **irretrievable**; sometimes getting temporarily or permanently stuck within locks.

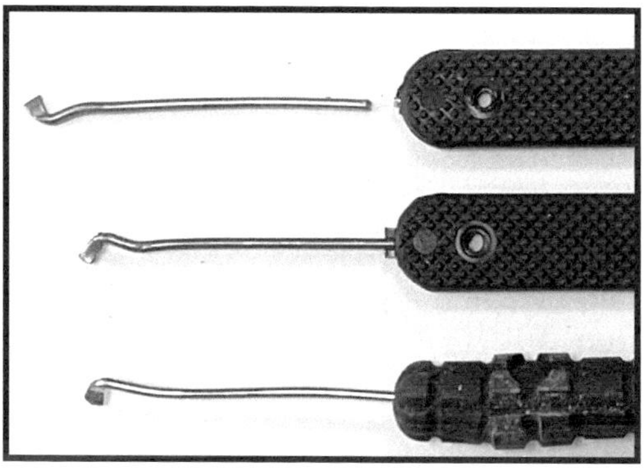

(Don't be surprised if your bypass drivers don't work as advertised. Don't get discouraged if they keep bending and breaking. This engineering flaw is known, and no tool manufacturer has solved this issue yet. But you can now use this information to be a step ahead with your planning and troubleshooting.)

PADLOCK SHACKLE SHIM

This is another technique that is not as straight forward as marketing and social media will lead you to believe. Most effective on the older and classic rotating dial combo locks that adorn most of the lockers in high schools across the country. An attack adjacent to the shackle.

- Insert your shim(s) into the body of the padlock, along the side of the shackle. While holding positive driving pressure with the shim, try to push and pull the shackle several times.
- (Most padlock shim tools come in multi-packs with shims of a variety of sizes. **These are not one-size-fits-all**.)

Sometimes called a "Padlock Shim" technique. Butterfly shim, and shackle shim are terms that are also used on occasion. A general shim technique is when you wedge any one tool in between two parts of a mehanism.

This is another technique that has a less than 100% chance of working, due to several variables. First, you need to know if this technique and tool will even work on the current padlock you have targeted. Is there enough of a gap that you can even fit the shim into the body of the padlock without damageing the tool? Will this tool rip and tear and get stuck within the lock (**irretrievable**)? Do you need only one shim for one side of the shackle or do you need two shims at the same time for both the heel and toe side of the shackle? Is your shim the right width and length (these shim sizes are not all created equal)? And yet again, manufacturers will often update the design of their locks to prevent well known bypasses from working (or at least not working as easily).

This is another technique that we don't recommend putting high up on your primary on scene options list, *unless* you have a clear indicator that this tool/technique is likely to work on the specific lock in question.

The odd shape and multiple sizes of shims required make this tool difficult and impracticle to carry in your 1st Line Gear, your EDC. For most emergency entries, you are likely to start your target assessment without access to this tool if it is carried in your 2nd Line Gear. So, remember to understand that almost every lock out there has multiple entry attack vectors, many different ways that you can get it to unlock.

This tool/technique combo is becoming less and less deployable as manufacturers change and update their lock designs to add a number of known preventative measures to stop this type of attack from working. Dial combo manufacturers, specifically Master Lock, has changed the shape of their internal mechanisms in such a way that internal components grab shim tools and prevent them from sliding the shackle open. And higher security padlocks often utilize a "ball bearing" locking mechanism, which cannot be opened with this tool/technique combo.

There are many different types of bypasses for padlocks, and other locks. And there are many different types of tools and techniques to implement different types of bypasses. Some bypass techniques can be categorized into types: kinetic bypasses (hitting the padlock with a hammer); bypass driver bypasses; shackle shim bypasses; furniture bypasses (manipulating what the lock is mounted to); and more. Different makeshift tools can often be used for many types of bypasses and padlocks. These are just some foundational options that you'll be able to identify and implement in the field; this chapter is certainly not all inclusive.

When you are building your skill set, you get to choose how to balance your preparedness. If your goal is *Tactical* Lock Picking, we can offer you two big guidelines to consider: 1) prepare your skill set for the *most common* types of locks that exist in your reigon, and 2) specifically the types of obstacles that your mission-role is likely to require you to interact with.

MIDDLE, Ga. (May 2024)

MIDDLE, Ga. (May 2024)

DOOR BYPASSES

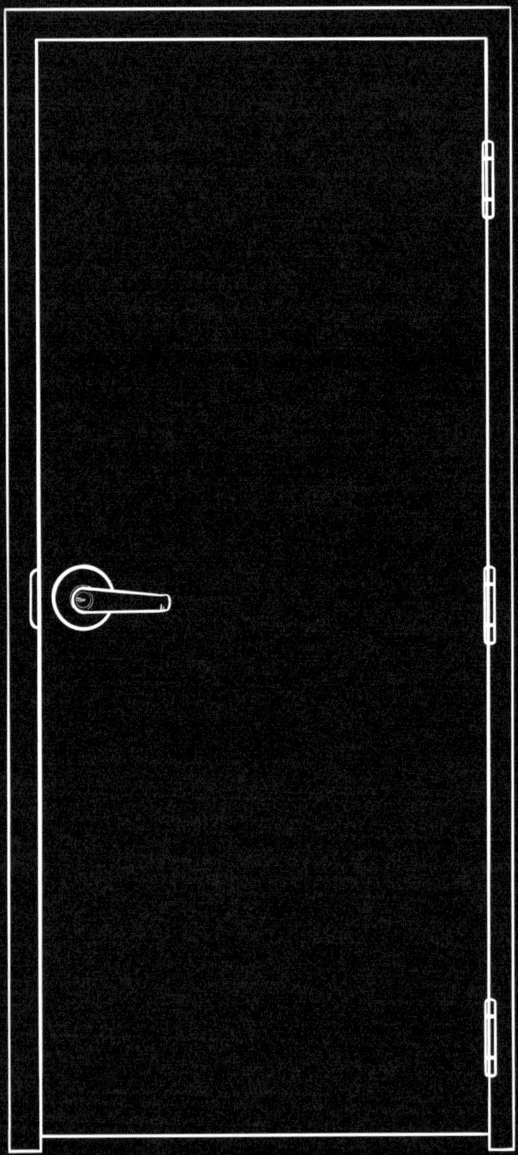

PULL DOORS

For modern conventional doors and doorframes, most "Pull Doors" (doors that you grab the knob/handle and *pull* towards you to open) are commonly set flush within the frame on the side facing the user. The door's latches are visible and therefore directly accessible between the door and the frame. When the door/frame gap is large enough, you can often gain either a little or a lot of information about the door's entry option potential. Because *pull door* latches are not set behind the spine that runs the length of the doorframe, the latches are usually directly accessible with a rigid tool, often able to be manipulated so that you can bypass the keyway and slip the latch to make entry through a locked door.

PUSH DOORS

"Push Doors," (where you would grab the handle/knob and *push* to open the door) are usually set back on the far side of the spine of the doorframe. This spine visibly and physically blocks access to the door's latches. For many push doors, flexible tools can manipulate a latch by sliding and snaking around the door and the frame to push up against the latch, causing the latch to recede into the door (as if the knob was turning and thus pulling the latch into the door itself).

The tools and techniques in this primer are not exclusive or exhaustive. Some of our favorite and most reliable tools for the pull-door side of a locked door are wire coat hangers (easy to source from many environments), or the Traveler's Hook seen below with the red handle (a tool that was designed for an entirely different use in an entirely different industry).

Tools created from scraps or garbage, tools that are custom designed, or tools that are professionally designed to rigorous specifications all have their own pros and cons.

It is important for you to experiment with different tools/techniques, but the real value of Tactical Lock Picking is that you should be paying attention to how you **troubleshoot** the process when a tool & technique is struggling or failing to help you gain entry.

Tools & Techniques are great...
...but problem solving is great-*er*!

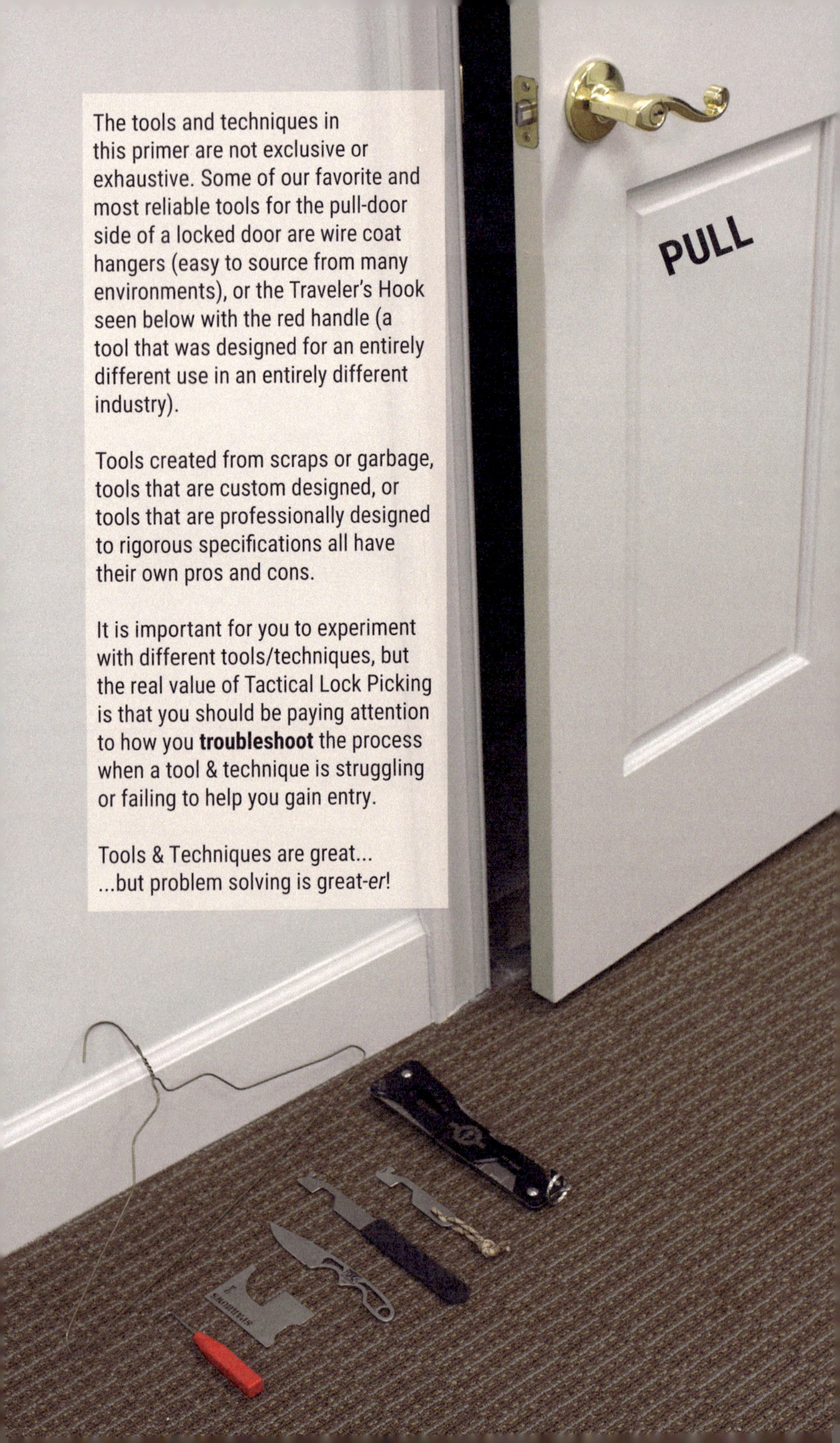

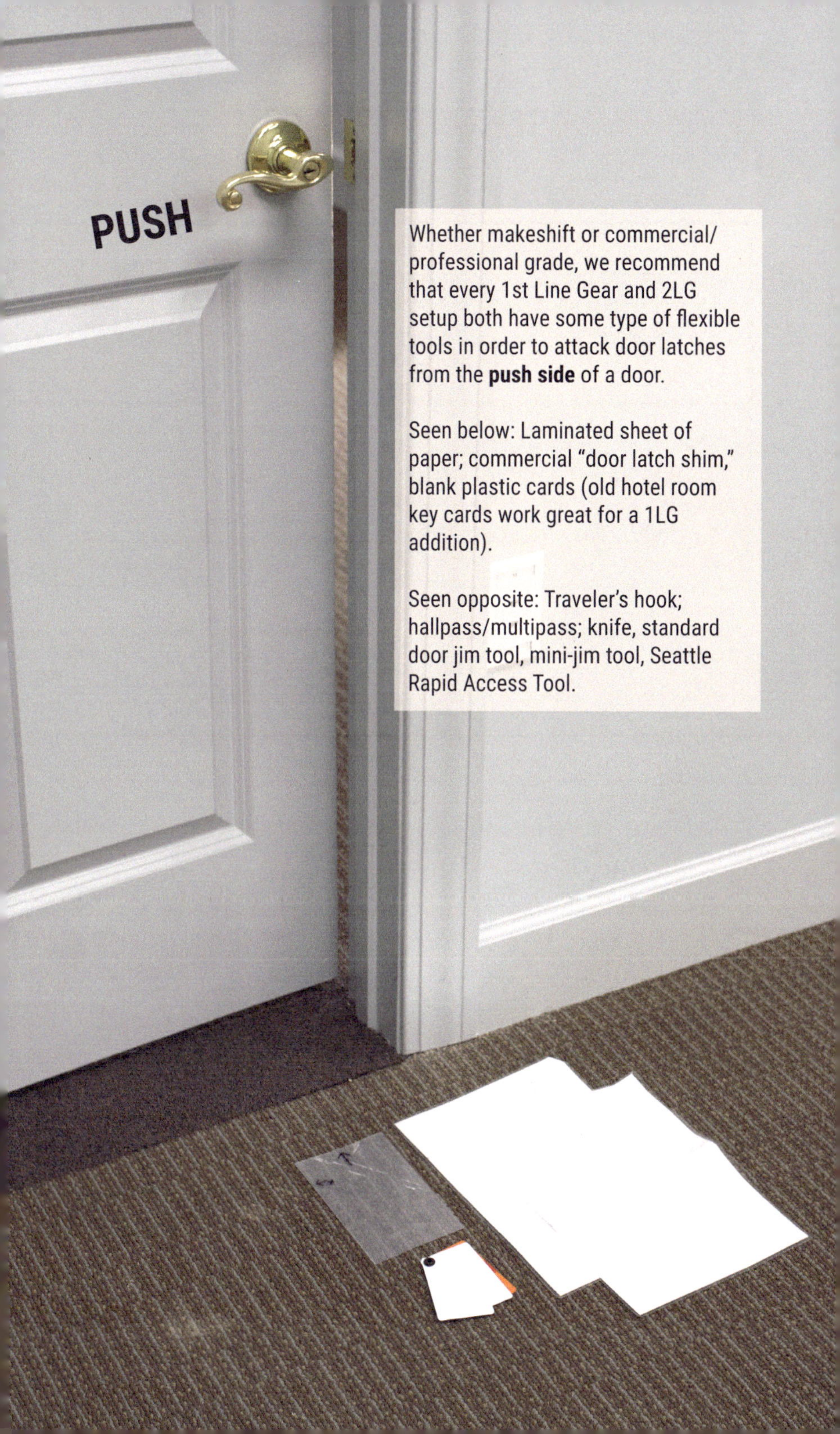

Whether makeshift or commercial/professional grade, we recommend that every 1st Line Gear and 2LG setup both have some type of flexible tools in order to attack door latches from the **push side** of a door.

Seen below: Laminated sheet of paper; commercial "door latch shim," blank plastic cards (old hotel room key cards work great for a 1LG addition).

Seen opposite: Traveler's hook; hallpass/multipass; knife, standard door jim tool, mini-jim tool, Seattle Rapid Access Tool.

SECURITY LATCH

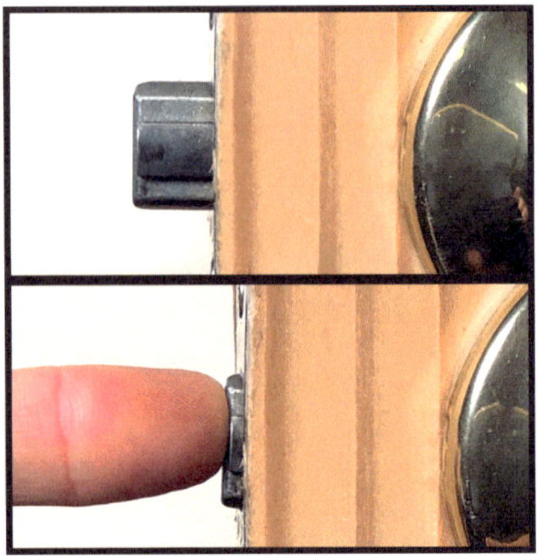

When the latch and the deadlatch mechanism are at rest, the latch can be pushed into the door.

When the deadlatch mechanism is pushed in by itself, then the latch is prevented from being manipulated.

When properly installed, the door should shut, the latch should extend into the latch strike plate, and the deadlatch mechanism should remain pushed into the door.

Sometimes, you can force the deadlatch mechanism to slip out of place and extend to its rest position - rendering the deadlatch mechanism ineffective. Allowing the latch to now be manipulated.

LATCH BYPASSES

PULL DOOR. Latch visible and accessible in a direct linear fashion. A rigid tool can be used to slip the latch.

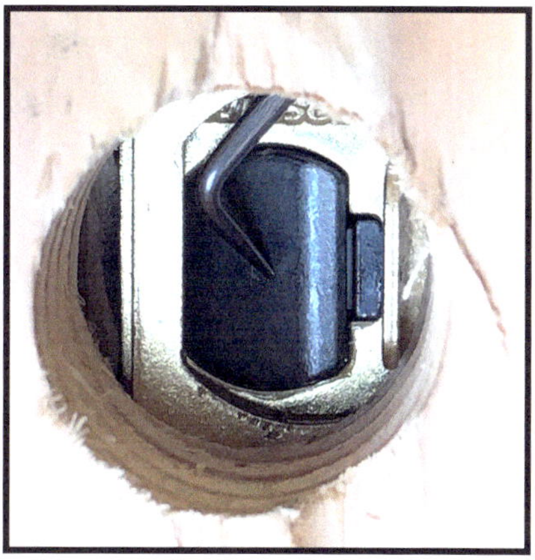

PULL DOOR. Side view (from within the door frame). The hook reaches behind the latch and rotates, compressing the latch, allowing the door to be pulled open.

PUSH DOOR. Latch is visually and physically protected by the spine of the door frame. A flexible tool is inserted.

PUSH DOOR. View from the side (within the door frame). A flexible shim weaves between the door and the frame and slides against the latch, compressing the latch, allowing the door to swing open.

DEADBOLTS

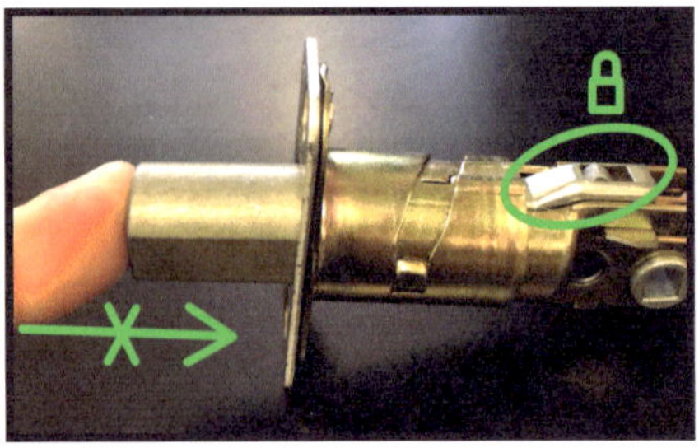

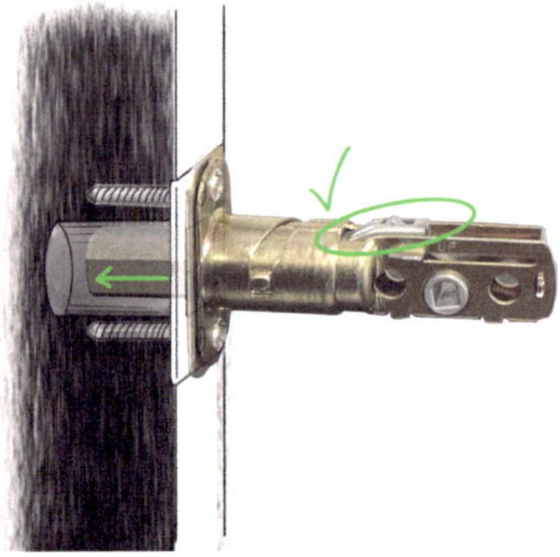

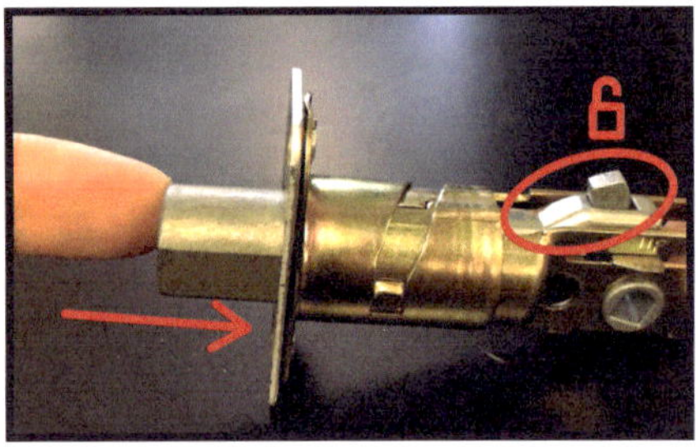

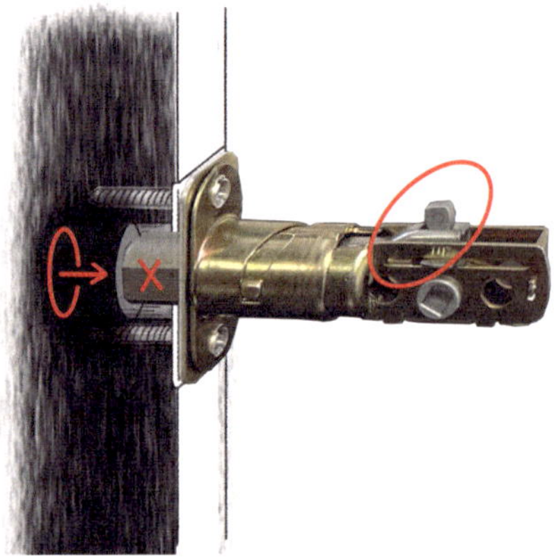

When properly installed, common residential deadbolts should extend 100% of their full extension range in order to become immovible, or "dead." If the receiving hole that is drilled into the door frame is too shallow, then even if you turn the deadbolt to the locked position, and you hear it go "thunk!" up against the wood inside the door frame, this does not mean that the deadbolt is fully extended. This *only* means that the deadbolt has extended until it has struck resistance.

When deadbolts are installed improperly on doors, more specifically - when the receiving holes are too shallow, deadbolts can often be bypassed by simply *walking* the deadbolt back into the door and into the unlocked position (next page).

DEADBOLT WALKING

If you can physically access the deadbolt (usually from the pull side of the door), use a sharp object **with care** and push down and "bite" into the deadbolt close to the door frame. Then slide or scoop your tool, and the deadbolt, in towards the door.

Then reset your tool and repeat, walking the bolt in towards the door in little steps until you achieve an opening.

Any time you're using a sharp tool and a lot of pressure **USE CAUTION**. This technique is usually only available from the PULL SIDE of a door, and the more narrow and tight the gap is between the door and the frame, the more difficult this technique is. And remember that this usually won't work if the deadbolt is properly installed and fully extended.

Lock picking (the act of using a tension wrench and a rake/or pick) is the first skill and the foundational skill we teach because it is such a great option to fall back on for when efficient bypasses, like deadbolt walking, aren't available. Remember that Tactical Lock Picking is *the process of solving access problems*, it's NOT always just about *picking a lock*.

UNDER DOOR TOOL (UDT)

Another very simple but not always very easy attack vector. Sometimes it will work perfectly, like magic, on one door. But experienced practitioners with a large number of data points to their field-studies, know that sometimes this technique can be a nightmare of troubleshooting to get just the right application that results in an entry.

The first thing that you need to know is that **this technique doesn't work on *every* handle**.

Just because you see a handle doesn't mean your UDT will solve the problem. We organize handle operations into three types when it comes to UDT's: Dead, Alive and Hybrid.

DEAD HANDLES

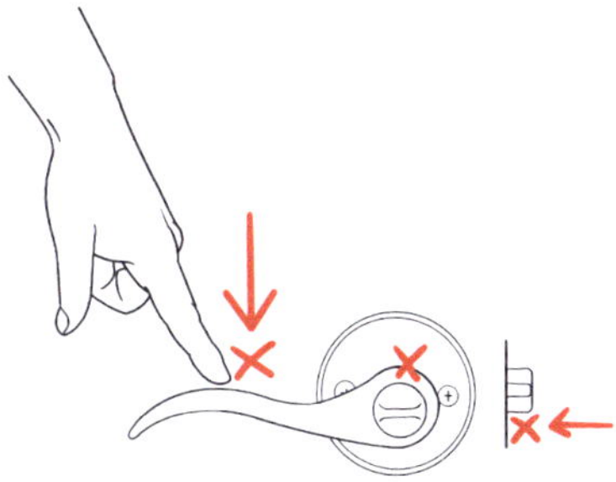

Dead handles usually have a "twist to lock" feature on the interior side of the door handle. Dead handles are the types of handles that when they're locked, if you grab the handle and try to twist the handle to the open position, the handle doesn't move, the lock doesn't unlock, and the latch doesn't move. The handle is just that, "dead."

ALIVE HANDLES

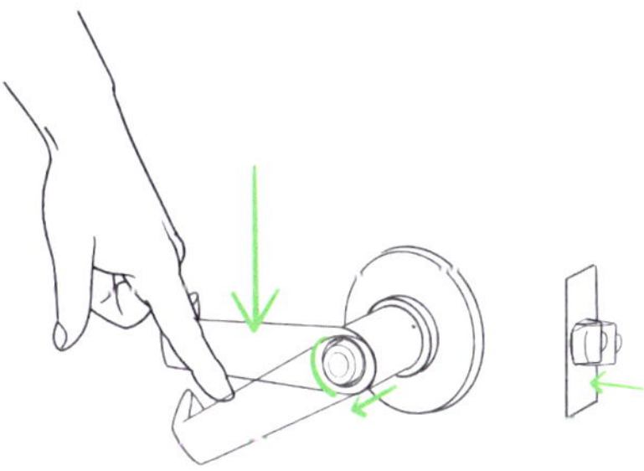

Alive handles usually have a push button lock on the interior side of the door handle. For most of these, when the handle is locked, and you rotate the handle on the inside of the room, the handle *does* rotate, the push button lock usually pops out to the *unlocked* position, and the *latch* usually pulls and recedes into the door, allowing the door to swing open.

HYBRID HANDLES

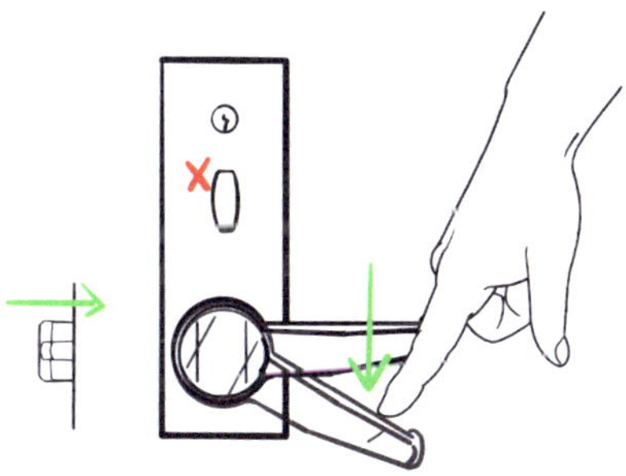

Hybrid handles are commonly found on hotel room doors, and on the commercial & government doors that use the "Kaba" style locks. These handles always allow people on the inside to twist the handle, which recedes the latch, allowing the door to open, but the locks *always* stay locked on the outside.

Visualization is the name of the game here. Going slow and steady will pay off in dividends because every small movement in your hands is a large movement at the end of the tool. Using your hearing and your visualization will help immensely.

A door with a large window allows you to see the working end of your tool and helps you adjust in real time. Having a friend look through a side window or back door and to give you advice, can also help. Even if you are operating solo, if you see a handle on the outside of a door, it's almost always worth it to try to walk around the structure to try to see if you can find a window somewhere that allows you to visually check the interior handle of your target door (and maybe the window might be open too!). Sliding a cellphone halfway under the door and using its camera so that you can see the interior door handle from below is a trick that we very commonly utilize and that we've been teaching for years.

I've never heard a report of the metal part of a UDT breaking. But, drawstrings *will* break during trainig and during operations. Part of your professional loadout shouldn't just be to have replacement cables available but also to have the tools and the practice to ensure that you know *how* to replace them during an emergency (a quality multi-tool and/or set of pliers is key for this).

"Beware bags, boxes, and bedframes," is a phrase that we started implementing in our curriculum in 2016. This helps practitioners cement the fact that UDT tools are another tool in the category of **irretrievables**, tools that have a likelihood of becoming stuck within locked obstacles. If the end or your tool snags a bag, or a pile of loose clothing, then you may not be able to apply any amount of dexterity to un-snag the end of your tool. It may end up being forever stuck no matter how hard you pull. Things like bed legs and table legs can also end up permanently or semi-permanently capturing the working end of your UDT.

Be cautions of your drawstring slipping and running up through the side of the door, between the door and the frame, usually this will hamper your attempts but on occasion it can apply some intentional force in a unique direction.

Beware the false confidence of having successfully used a UDT on an easy door handle *once*, gaining a quick entry, then thinking that you have a magic key that will always work. Try to find at least one example of the three types of door handles we categorized (alive, dead, hybrid) in order to get your feet under you and to have actually "earned" a solid foundation to stand upon for whether this technique will even work or not.

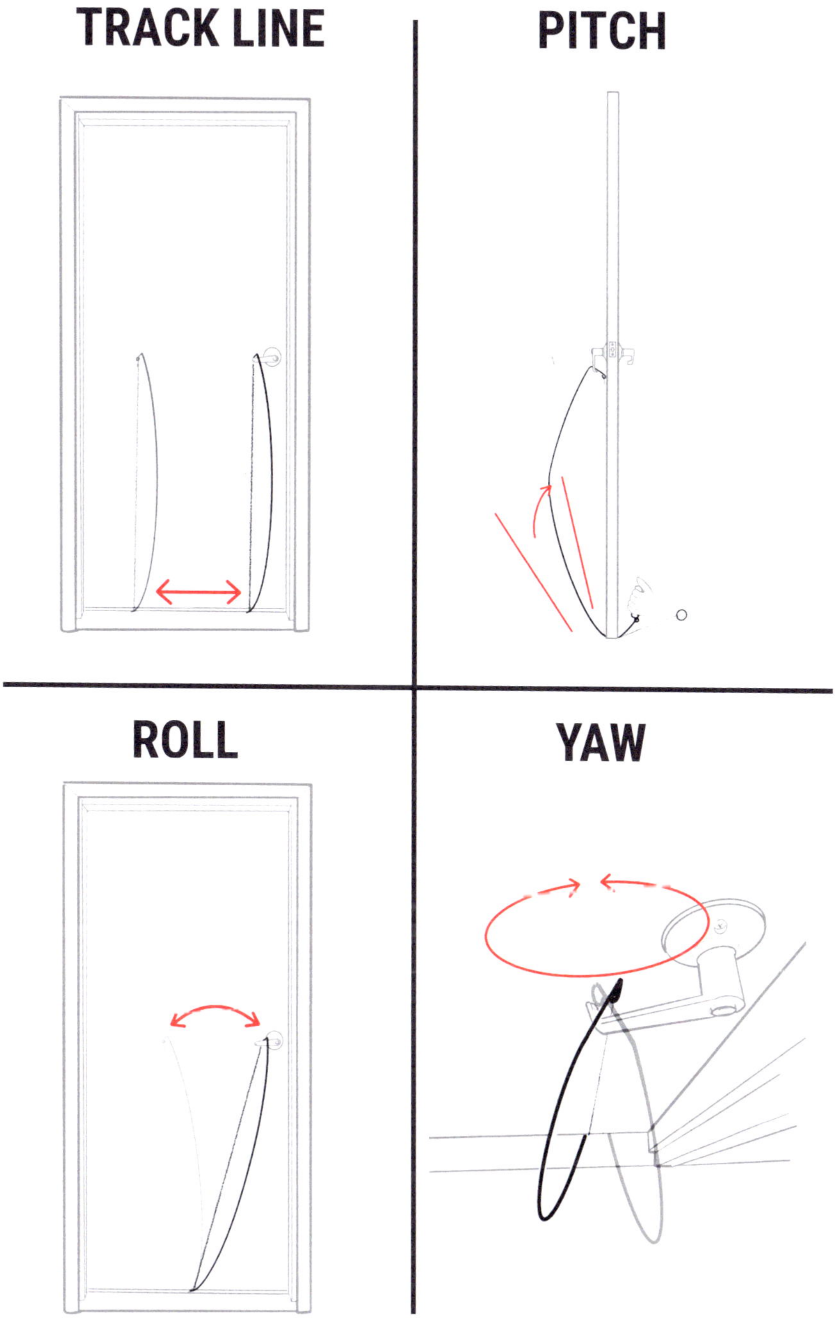

These movement and position concepts
can help you make the most of a UDT entry.

LISHI 2-IN-1 DECODERS

In the last ten years, the Lishi "Two-in-one Picks/Decoders" have had a huge impact on the field of non-destructive entry (NDE). Lishi used to be known for their vehicle keyway entry tools, mostly wafer locks. When they moved into the pin-tumbler world for residential as well as commercial keyway tools, they earned a ton of popularity, and rightfully so.

These have made Single Pin Picking (SPP) entry attempts much more widely accessible and easily achievable for beginners and experts alike.

But(!) they are still not as simple as "plug and play..."

KW1 MODEL

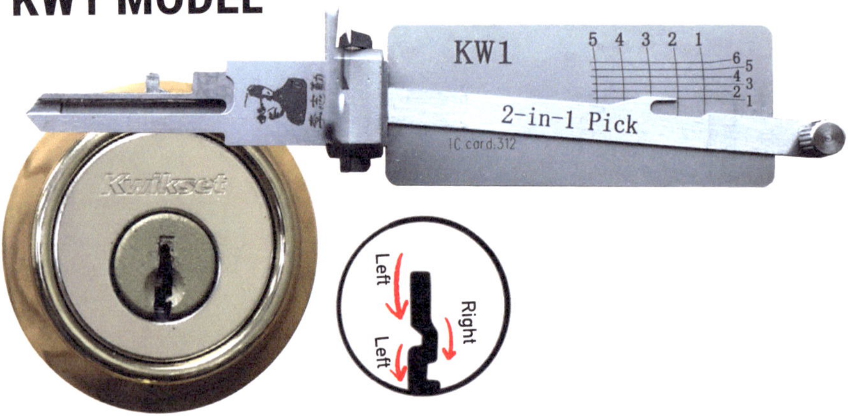

SC1 MODEL

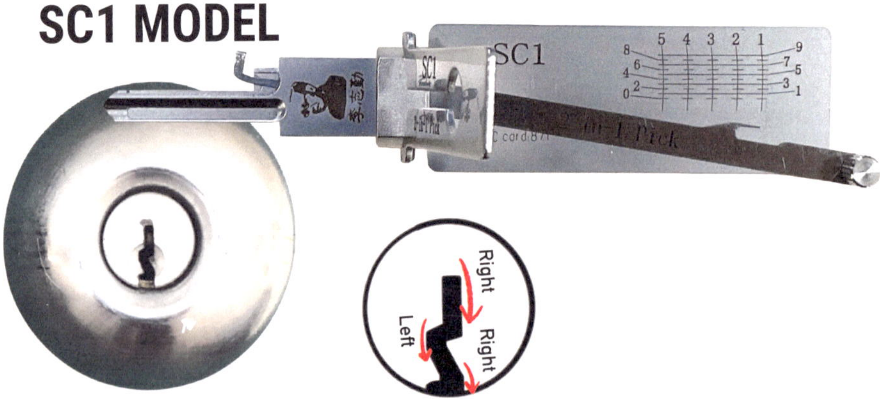

- You do not need to search very hard to find creators on social media that will make Lishi tools (and many other NDE techniques) look like they are magic. Many will make their skill application look as if they never struggle and never "miss a shot." In the *real world*, lock entry is *always* a percentage game of what might work, and how to problem solve it when it doesn't.
- Lishi decoders require quite a bit of tradeoff between their monetary cost and the amount of space they take up. These things are somewhat bulky, so whatever space they take up in your bag can't be used by other entry tools.
- They do occasionally require maintenance. These picks can bend and the hex-screws can loosen on these tools, requiring replacement or repair.
- With these tools, sometimes you will have very smooth motion applying tension and picking in one direction but will have very sticky and scratchy progress applying tension in the other direction. With practice and creativity, and solid training courses focused on field application, you can use this to your advantage.
- This method doesn't take a LOT of homework, but does take SOME homework. Make/model recognition is incredibly important for emergency and operational field use.
- If you are excited about Lishis but you have a limited budget and you are still firmly within the initial learning phase of your NDE journey, we very highly recommend only buying these two Lishi tools seen opposite, the Lishi KW1 and the Lishi SC1.
- If you are a beginner, it doesn't matter too much if you get a Lishi that is "Left Handed" or "Right Handed (Standard)," they will both work the same. All this means is which side the bitting gauge chart is on and which side the tension handle is on. This makes *almost* no real difference in the application of your craft except for in extremely rare circumstances.

If you don't already own a pin-tumbler Lishi, and haven't already achieved a base level reliable and repeatable capability with them, then also don't worry about buying more than just the standard two Lishis's that we recommend (shown: opposite).

If you're still very new to this skill then we recommend not worrying about using a 6-pin decoder for a 5-pin lock and having to figure out the adjusted spacing or having to purchase and carry a manual spacer for the end of your tool. If you want to disregard this recommendation because buying more tools makes you happy, then we whole heartedly support you. Our recommendations here are for the self-supported operator with limited time/money/and infrastructure.

The best thing that we can leave you with is that these Lishi tools and techniques (like all the others we teach) are not magic. Do not get discouraged if you struggle to get them to work consistently and smoothly right away. There are world class experts that still struggle on locks from time to time (yes, even when they use Lishi tools).

LISHI OPERATIONS

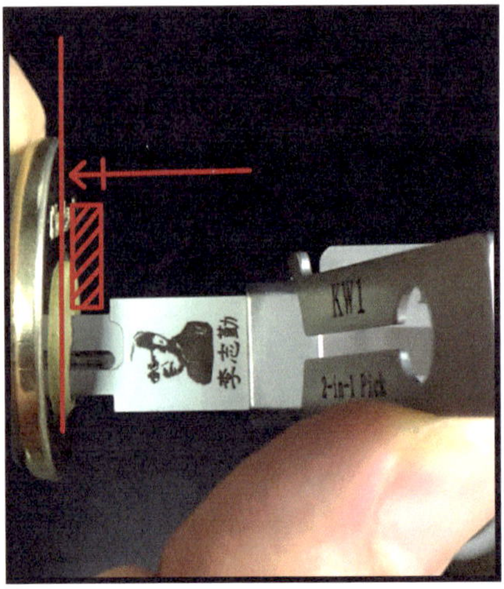

- Ensure that your Lishi tool is inserted all the way flush up against the keyway's face, to the shoulder of the Lishi tool (photo: left, shows partial insertion). Sometimes this can be tricky and can require you to move your lifter-pick up and down a little during the insertion process. Reassess throughout your entry attempt to ensure that your Lishi is *always* flush with the lock face.

- Open the swing arm and provide turning tension in your best guess at the correct direction for tension. Start applying light steady tension in one direction.
- Keeping the handle/knob end of the pick high, gently tap downwards on each of the pin stacks (using the chart on the Lishi. Photo below), if the pins have even a little bit of springy-ness then leave them alone. You should be looking for the pins that are stiff, or "binding," and you should gently try to lift and to "set" those binding pins.
- Once you hear, see, or feel a binding pin that is set into place, leave that one alone and go look for the next binding pin.
- If you find a binding pin and are struggling to lift it, it is likely because you are applying too much turning tension with the swing arm, while simultaneously trying to lift the pin. When lifting a binding pin, remember to ease up on your tension a little bit so that you are not working against yourself.
- Each lock has a different personality. It doesn't necessarily matter which direction you progress from pin stack to pin stack, (front to back, or back to front). You may have to try both of these options, as well as a mixed option, as you troubleshoot each lock.

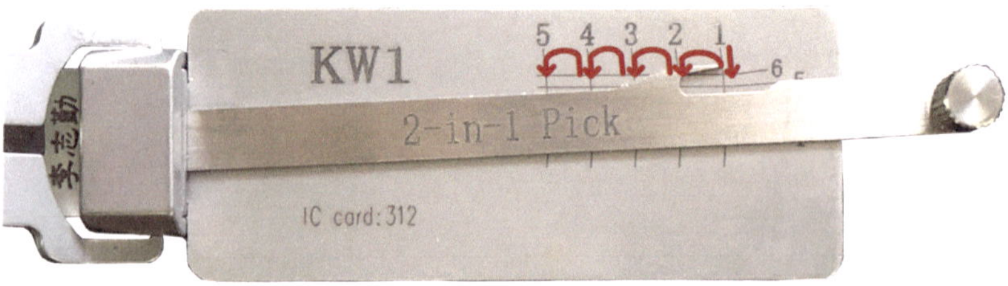

KW1
5 4 3 2 1
6
5
2-in-1 Pick
C cord.312

COMB TOOLS

The UTAC Comb Card Pack, and the associated locks they *should* open. Other makes/models (especially "clones") are often succeptable to this type of attack as well.

Every type of comb tool has its own pros and cons. What's important for field operators to know is how to quickly identify locks and to know which specific comb will open your targeted lock. This doesn't take *too much* research/memorization, just a little. It's important to remember that comb tools belong to a classification of tools that UTAC calls **irretrivable** tools; these have a *known* potential of becoming stuck within a locked obstacle & can be difficult or impossible to remove. Don't ever jam a comb tool into a lock, these combs should all insert fairly easily. If you have to struggle to get a comb tool *into* a lock, don't be surprised if you struggle or fail to *remove* it.

Our **6-pin comb** (Opposite, top left) should open many of the Master Lock 6-pin locks, especially the puck locks. Exceptions always exist.

The **6-pin Thin* comb** (Opposite, top right) should open the American Lock A748 padlock. We highly recommend using a tension wrench in conjunction with this comb attack. This thin comb may bend or break due to the turning force and spring tension required to rotate this keyway.

The **5-pin comb**, (Opposite, middle) should open the Master Lock 150, the Master Lock 160, and the Abus 55/60 padlocks seen in the middle row, opposite.

The **4-pin comb** (Opposite, bottom row) is fantastic at opening the Master Lock 140's and 141's.

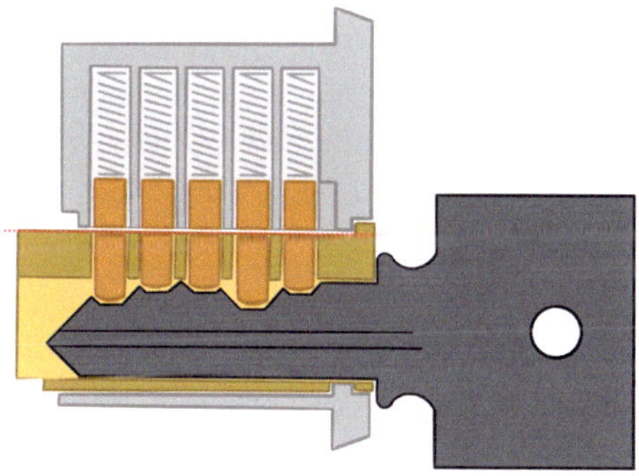

(Left) A lock with the proper key inserted will raise the top of the key pins to the shear line, and raise the driver pins above the shear line. When this shear line is clear, there are no obstructions preventing the plug from spinning = unlocked and able to spin.

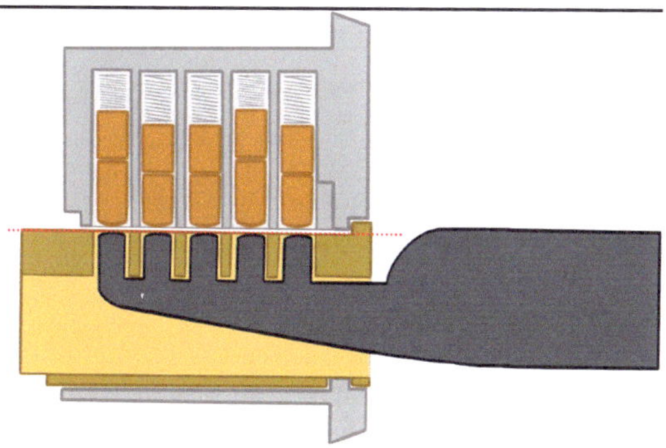

(Right) A lock opened with an appropriate comb tool. This lifts the driver pins *and* the key pins up above the shear line, removing all obstructions that keep the plug from spinning.

Just like every other tool, **comb tools are not magic**. For a comb tool to work on a lock, you need *several* factors to all line up correctly.

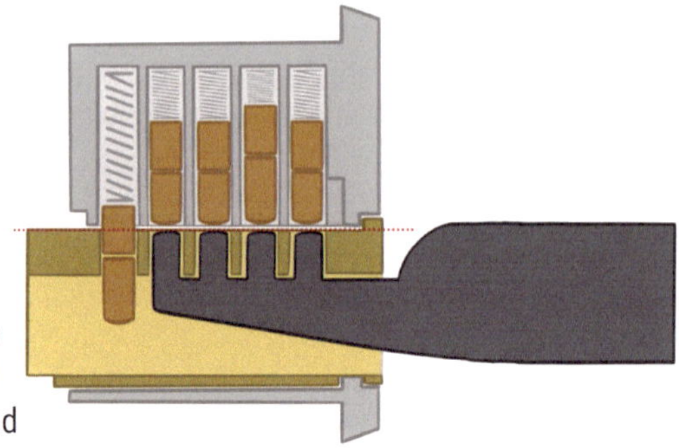

Up first, the comb tool should have the same number of lifters as there are pin stacks within the lock. (Right) A four-pin comb will not open a five-pin stack lock. And a six-pin comb may not open a five- or a four-pin lock. (There are always exceptions)

Second, the spacing of the lifters must match the spacing of the pin stacks within the lock. (Left) If the comb lifters are too closely or too widely spaced, the comb tool will not be able to fit into the chambers properly and will not be able to lift the key pins above the shear line.

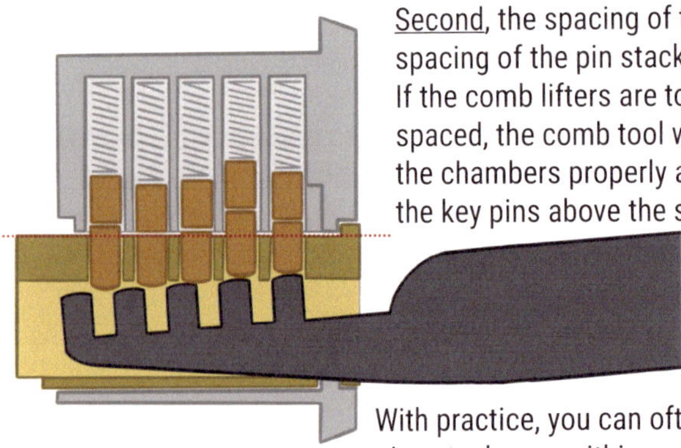

With practice, you can often figure out how many pins stacks are within an unfamiliar lock, but it's very difficult in the field to try to figure out if the *spacing* of the pin stacks is the same as the comb tool you have available.

One of the first difficulties that students have when they first try this technique, is lining up the comb lifters with the pin stacks and then lifting them *straight* up without the comb tool skipping off deeper or more shallow than intended. This is something to be aware of during your troubleshooting, and it just takes practice.

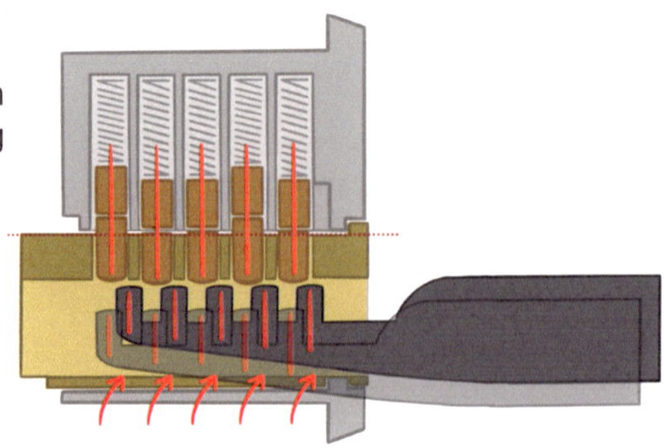

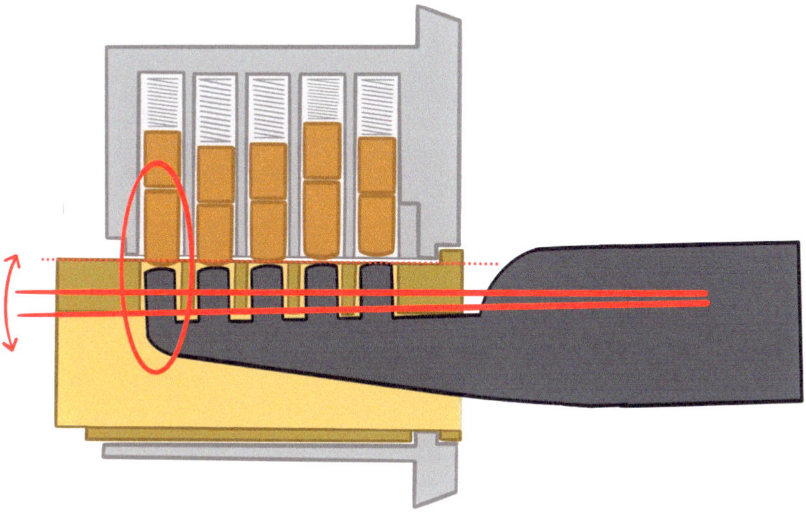

The most common problem that we see with students is that they can have every single variable correct but because of the spring tension of the pins, the comb lifters closest to the handle *are* in fact lifting the pin stacks fully above the shear line, but the shear lines of the deeper pin stacks within the lock are resting slightly too low, preventing the plug from spinning (shown above). Be sure that your tool stays flat when you are lifting the pin stacks.

Our recommended field usage of combs has two main components.

1) Because of speed and percentage of likelihood of successful entry, *if it is very clear* that a lock is combable, and very clear that you currently have access to the right comb for the right lock, combing is often a *first* attack vector.

2) Because of the possiblity of a comb tool becoming *irretrievable*, if the make/model of lock and comb tool is *not very clear* whether it will work or not, it's often the very *last* method on the list. A tool becoming irretrieveable can slow down an entry during an emergency or an operation, and it can even prevent many other entry options from being attempted.

Everything within the field application of Tactical Lock Picking a percentage game. Every tool & technique has pros and cons. Every single operator will have different skills. There are no two operators that will make the same exact NDE entry, even with the same tools and the same training. The *artform* of the application of NDE is unique to every operator.

COMBO LOCKS - PHYSICAL DECODING

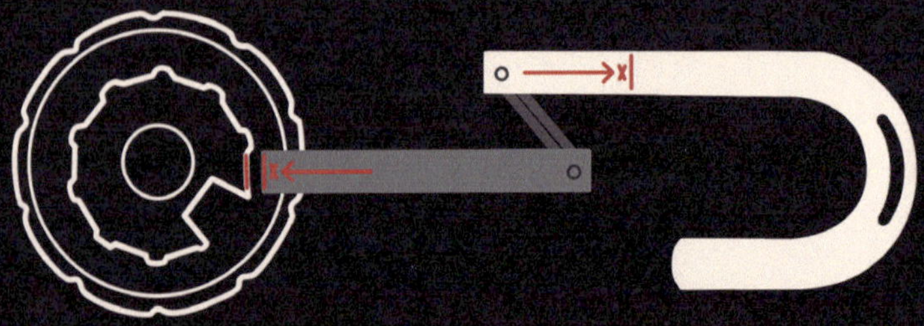

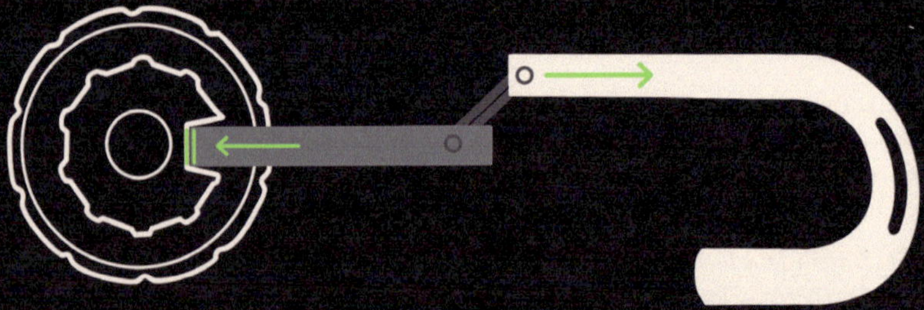

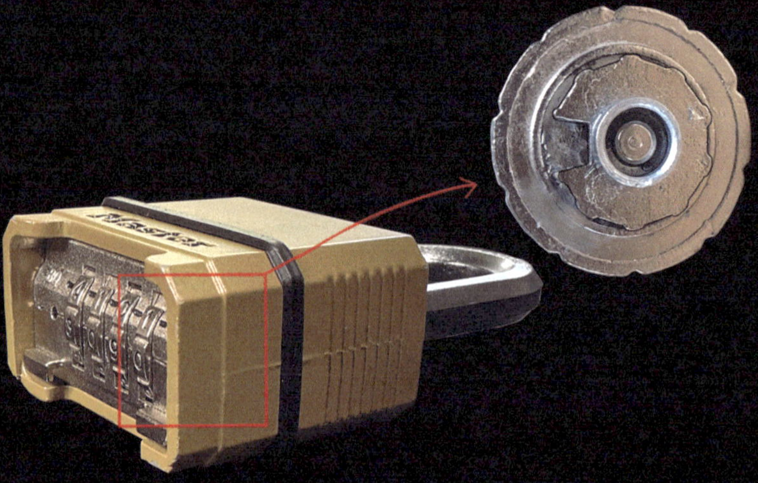

This technique does not work on every make/model of combination lock. You must have enough room adjacent to the wheels in order to slide your decoding tool into the body of the lock. The lock also needs to be free of shielding that would prevent your tool from reaching the inner wheels/gates. You also need to determine which side of each wheel the gates are on.

Step 1: Isolate one wheel to work on. Try to identify one spot within the inner wheel (one of the ten positions) that is noticably different, with a notched "gate" cut out of the wheel. Once you locate what you think is the gate, leave that gate centered in its current position.

Step 2: Identify the gates in the remaining wheels, and align all of the wheels so that all of their gates are perfectly aligned with one another (shown below).

Step 3: Add one digit to each of the wheels. Then try the shackle. Rotate all of the wheels +1 again, then try the shackle, and repeat until opening. If the shackle doesn't open within ten position attempts, you'll need to start troubleshooting.

Most locks have a 180-degree offset (meaning you could just add 5 to all the wheels to begin with) but if the offset is anything other than 180-degrees then our "add 1" method will guarantee that you find that offset.

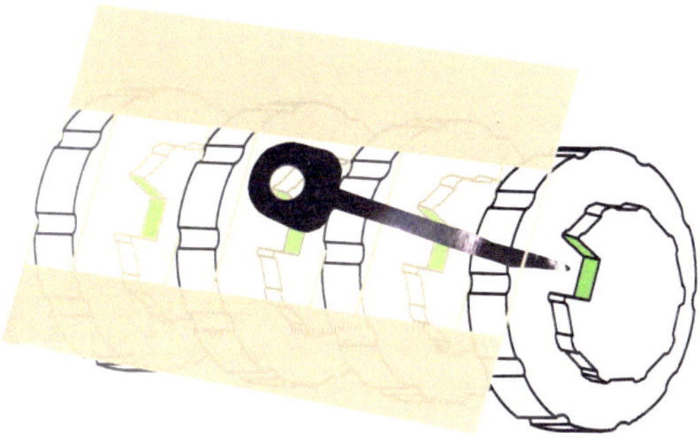

Troubleshooting is the part that truly shows the skills of a certified Tactical Lock Picking operator.

Often students will locate a gate for the facing positions on each of the wheels they are decoding, but there may be a wheel or two that are off by one digit; meaning that they can feel *a* click on a specific wheel, but they've actually just located the edge of the gate, and not the full gate facing in the centered position. Be sure all the gates line up *perfectly*.

New students of ours will also frequently have all four gates perfectly aligned on the facing side of the lock, but when they rotate them to the interior of the lock, they rotate wrong; either "fat fingering" it, or just counting wrong.

It's a good idea to document where you think the gates are showing at, for example, writing down "Gates located at 7-5-7-5," so that if you make a mistake while rotating the gates, you don't have to go all the way back to the beginning to find the gates again from scratch.

COMBO LOCKS
SOCIAL DE-ENGINEERING

There is a field of study and application called Social Engineering. Loosely defined, it is the act of getting a person(s) to take an action that you want them to take; ie having them open or unlock a door for you. For the Tactical Lock Picking curriculum that we have created, it is unlikely that responders will need to apply emergency or operational services behind a locked obstacle while also having someone present with the key or code. In such case, there would be no need to trick or convince the keyholder to unlock the obstacle for us.

TLP does, however, include what we call Social De-Engineering; understanding the common patterns of how humans interact with their own locked systems, and using those known patterns in order to achieve our own entry (when someone is not present with the correct key or code to assist us).

You can probably achieve multiple TLP Social De-Engineering entries within the next 24 hours if you keep your eyes open. Every single combination lock holding a vehicle gate shut, every single realtor key lock box on the front or side of a home, every single digital keypad used for a digital door lock; if you knew some simple information about each of the facilities, or about the people that set the combos for the locks, a large percentage of combinations chosen would have some type of associated meaning to the numbers selected for the code. Easily guessable. In your own life you are probably using three or four-digit codes right now that have easily guessable associations with information about your own life.

The act of being able to recognize common data and number patterns and to guess someone's combo, in and of itself, is a party trick at best. *The act of utilizing this Social De-Engineering attack vector (or any vector), as a part of a whole system of application...* **that is what makes an expert entry operator**.

What you can start doing today is gathering what we refer to as a blanket master list of combination codes. At its most base definition, the blanket system is simply opening a text file on your smart phone where any time you learn of a four-digit entry code, you simply write it on the list and then move on with your life. At its advanced level, the blanket master list can be utilized as a master data list, if you choose to do the homework and to organize these codes in order and to document frequency of codes, and to assign the types of associated data that pairs with the code (if any).

You can be as simple or as advanced as you want, but the act of looking at and documenting codes will be a positive step towards a life-long tool that you can leverage in nondestructive entry operations.

BRUTE FORCE CODE ATTACKS PUSH BUTTON LOCKS

1. 1234
2. 1243
3. 1324
4. 1342
5. 1423
6. 1432
7. 2134
8. 2143
9. 2314
10. 2341
11. 2413
12. 2431
13. 3124
14. 3142
14. 3214
16. 3241
17. 3412
18. 3421
19. 4123
20. 4132
21. 4213
22. 4231
23. 4312
24. 4321

Understanding the ins and outs of all the different paths through which you can decode, decypher, and otherwise unlock 3- and 4-digit codes can be an entire world of study. The push button style combination locks are particularly unique in that they are usually very easy to exploit using grade-school level math and logic.

In the photo opposite is a standard 10-digit array of inputs where the 1st, 2nd, 3rd and 4th input are selectd. On a *digital* keypad, if the first four inputs are worn (making them the most likely four digits for the access code) then there are 24 possible ways to arrange those four separate inputs into specific orders.

On a push button *mechanical* lock, if the selection of those four worn inputs makes up the access code, then the order in which you select those buttons usually does not matter. If you know which 4 buttons are used, then you don't have to try potentially 24 different combinations, but only one combination since, with most push button locks, the buttons just have to be selected as a group and in no specific order. This severely reduces the numbers of combinations required on a combo discovery journey.

Even more of a helpful factor in decoding push-button locks is that usually you cannot repeat digits in a code. Meaning that inputting 2-2-2-5 into a four-digit push button lock would be an invalid attempt since it only represents the two inputs of 2 and 5. If numbers cannot be repeated in a code then there are even less potential codes that you would have to try, when searching for the access code.

If you have the will to learn and to see the patterns and any interest in working through the simple math, then you can make these boiled-down code lists yourself with pen and paper, or you can even use simple number generator websites online. But(!) even if you are not a fan of math, there are still ways that you can exploit these push button locks by letting someone else (us) create these lists for you.

I carried an expensive and bulky lock pick gun (aka snap gun) in my 2nd Line Gear bag for almost a decade and used it zero times in the field. After the UTAC team designed our Brute Force Notebook (from scratch) I made the decision to remove that pick gun from my 2LG and I instead put our BFN into its place within my bag; and 3 years after that change I still have no regrets, this was definitely a valuable gear line adjustment.

We have an entire Brute Force Notebook available that is designed for field use. Its paper texture is non-glossy making it easier to write on, the paper is thinner allowing for more pages with a lower cost, it is spiral bound for ease of writing on it and reading from it in field conditions, and each code section has multiples of lists so that if you use one of the numbers lists you can tear it out and there is a new fresh list behind it ready to go. For *this* book, we wanted to introduce you to this technique and to teach you a little about how it works.

10-INPUT PUSH BUTTON LOCKS

**and similar locks following the same rules*

For a mechanical push button combination lock that follows these rules...

- There are **10 inputs** to choose from...
- The combination is **4 digits** long...
- The input **order does not matter**...
- And inputs **cannot repeat**...

...there are only **210 possible combinations** you would have to try, in order to try every possible code for that lock. Assuming no other errors and no exceptions exist (broken lock, a code length other than four-digits, a lock with no code installed, operator error, etc.) this should take you **around 10 minutes**, maximum.

1234	1260	1360	1490	2345	2456	2570	3469	3689	4680
1235	1278	1378	1567	2346	2457	2589	3460	3680	4690
1236	1279	1379	1568	2347	2458	2580	3478	3690	4789
1237	1270	1370	1569	2348	2459	2590	3479	3789	4780
1238	1289	1389	1560	2349	2450	2678	3470	3780	4790
1239	1280	1380	1578	2340	2467	2679	3489	3790	4890
1230	1290	1390	1579	2356	2468	2670	3480	3890	5678
1245	1345	1456	1570	2357	2469	2689	3490	4567	5679
1246	1346	1457	1589	2358	2460	2680	3567	4568	5670
1247	1347	1458	1580	2359	2478	2690	3568	4569	5689
1248	1348	1459	1590	2350	2479	2789	3569	4560	5680
1249	1349	1450	1678	2367	2470	2780	3560	4578	5690
1240	1340	1467	1679	2368	2489	2790	3578	4579	5789
1256	1356	1468	1670	2369	2480	2890	3579	4570	5780
1257	1357	1469	1689	2360	2490	3456	3570	4589	5790
1258	1358	1460	1680	2378	2567	3457	3589	4580	5890
1259	1359	1478	1690	2379	2568	3458	3580	4590	6789
1250	1350	1479	1789	2370	2569	3459	3590	4678	6780
1267	1367	1470	1780	2389	2560	3450	3678	4679	6790
1268	1368	1489	1790	2380	2578	3467	3679	4670	6890
1269	1369	1480	1890	2390	2579	3468	3670	4689	7890
									-END-

12-INPUT PUSH BUTTON LOCKS

and similar locks following the same rules

For a mechanical push button combination lock that follows these rules...

- There are **12 inputs** to choose from...
- The combination is **4 digits** long...
- The input **order does not matter**...
- And inputs **cannot repeat**...

...there are only **495 possible combinations** you would have to try, in order to try every possible code for that lock. Assuming no other errors and no exceptions exist (broken lock, a code length other than four-digits, a lock with no code installed, operator error, etc.) this should take you **around 20 minutes**, maximum.

The list of possibilities on the next two pages for this particular make/model lock has been hand crafted by the UTAC team. You will notice, if you look at the previous lock's list, that the first 210 codes on the following page are the same as the previous page. This is intentional as most people that choose combinations for locks like the one pictured left (with 12 inputs) usually don't choose the star or the pound inputs, so we rearranged the list to put the numbers-only combos up at the front of the list and the combos containing symbols at the end of the list. We also were sure to put the classic "1-2-3-4" at the beginning of the list as it is an incredibly common code in the wild.

We also threw into the mix the fact that we made the zero inputs a ten-value instead of a value of less than one; this makes the flow of movement across the keypad much more natural and efficient (And if this doesn't make sense to you because you're not a "numbers person" then all it means is that we nerded out to bring you an ever smoother experience while inputting mass code lists).

This particular list, 495 entries in total, doesn't take too long to input so we recommend moving a little slower than you think is necessary in order to prevent accidents like fat-fingering a code, or not activating the unlock latch correctly, etc. Better to move slower and be more sure that you gain entry, than to move too fast and make a mistake and have to try the list again.

Also remember that a brute force code attack might not be the best method for entry! The UTAC TLP curriculum is *a framing of application* and not merely a selection of tools and techniques. You, as the entry specialist, must always take into account the entire wholistic operation and you should weight your entry options wisely.

In the real world, entry skills are never implemented merely for the sake of unlocking an obstacle. Skills are always implemented in order to achieve the *thing* on the other side of the locked obstacle. The best practitioners in the world in this skill set are not necessarily the most advanced at applying a tool and a technique; they are the ones that make the smartest *decisions* concerning the operation.

1234	1345	1469	2345	2469	3456	3689	5679	120*	140*
1235	1346	1460	2346	2460	3457	3680	5670	120#	140#
1236	1347	1478	2347	2478	3458	3690	5689	12*#	14*#
1237	1348	1479	2348	2479	3459	3789	5680	134*	156*
1238	1349	1470	2349	2470	3450	3780	5690	134#	156#
1239	1340	1489	2340	2489	3467	3790	5789	135*	157*
1230	1356	1480	2356	2480	3468	3890	5780	135#	157#
1245	1357	1490	2357	2490	3469	4567	5790	136*	158*
1246	1358	1567	2358	2567	3460	4568	5890	136#	158#
1247	1359	1568	2359	2568	3478	4569	6789	137*	159*
1248	1350	1569	2350	2569	3479	4560	6780	137#	159#
1249	1367	1560	2367	2560	3470	4578	6790	138*	150*
1240	1368	1578	2368	2578	3489	4579	6890	138#	150#
1256	1369	1579	2369	2579	3480	4570	7890	139*	15*#
1257	1360	1570	2360	2570	3490	4589	123*	139#	167*
1258	1378	1589	2378	2589	3567	4580	123#	130*	167#
1259	1379	1580	2379	2580	3568	4590	124*	130#	168*
1250	1370	1590	2370	2590	3569	4678	124#	13*#	168#
1267	1389	1678	2389	2678	3560	4679	125*	145*	169*
1268	1380	1679	2380	2679	3578	4670	125#	145#	169#
1269	1390	1670	2390	2670	3579	4689	126*	146*	160*
1260	1456	1689	2456	2689	3570	4680	126#	146#	160#
1278	1457	1680	2457	2680	3589	4690	127*	147*	16*#
1279	1458	1690	2458	2690	3580	4789	127#	147#	178*
1270	1459	1789	2459	2789	3590	4780	128*	148*	178#
1289	1450	1780	2450	2780	3678	4790	128#	148#	179*
1280	1467	1790	2467	2790	3679	4890	129*	149*	179#
1290	1468	1890	2468	2890	3670	5678	129#	149#	170*

170#	239#	258#	289*	356*	379*	450*	480#	589*	690#
17*#	230*	259*	289#	356#	379#	450#	48*#	589#	69*#
189*	230#	259#	280*	357*	370*	45*#	490*	580*	60*#
189#	23*#	250*	280#	357#	370#	467*	490#	580#	789*
180*	245*	250#	28*#	358*	37*#	467#	49*#	58*#	789#
180#	245#	25*#	290*	358#	389*	468*	40*#	590*	780*
18*#	246*	267*	290#	359*	389#	468#	567*	590#	780#
190*	246#	267#	29*#	359#	380*	469*	567#	59*#	78*#
190#	247*	268*	20*#	350*	380#	469#	568*	50*#	790*
19*#	247#	268#	345*	350#	38*#	460*	568#	678*	790#
10*#	248*	269*	345#	35*#	390*	460#	569*	678#	79*#
234*	248#	269#	346*	367*	390#	46*#	569#	679*	70*#
234#	249*	260*	346#	367#	39*#	478*	560*	679#	890*
235*	249#	260#	347*	368*	30*#	478#	560#	670*	890#
235#	240*	26*#	347#	368#	456*	479*	56*#	670#	89*#
236*	240#	278*	348*	369*	456#	479#	578*	67*#	80*#
236#	24*#	278#	348#	369#	457*	470*	578#	689*	90*#
237*	256*	279*	349*	360*	457#	470#	579*	689#	-END-
237#	256#	279#	349#	360#	458*	47*#	579#	680*	
238*	257*	270*	340*	36*#	458#	489*	570*	680#	
238#	257#	270#	340#	378*	459*	489#	570#	68*#	
239*	258*	27*#	34*#	378#	459#	480*	57*#	690*	

5-INPUT VARIABLE LOCKS

If you are looking at a mechanical push button combination lock and it has this 5 push-button vertical arrangement in this classic brushed metal, or the smaller round 5 black button circular arrangement, you're probably looking at a Kaba Simplex type lock. These are incredibly common on government buildings, and large corporate buildings, or service-related buildings.

The math is pretty complex but there are only **1,081 possible combinations** for these 5-button Kaba Simplex type locks. At a fast speed, assuming no other errors and *no exceptions exist* (like a broken lock, or a stubborn door fitment, or a lock with no combo installed, etc.) it could take you roughly **1 hour and 45 minutes** to input every possible code.

These unique locks allow for multiple buttons to be pushed together at the same time as a unique input. **For the following code list**, standard numbers should be a single press input, and numbers that are in parentheses together should be pressed together at the same time. The code of 12345 on the following pages would be inputted by using just one finger and pressing the 1, then pressing the 2, etc. But, the code of 1(23)(45) would be inputted by pressing the 1 input, then pressing both the 2 and 3 together at the same time, then pressing the 4 and 5 together at the same time. And so on.

The instructions for this lock direct the user to input the code of (24)3 as the first test code, to test the locks operation upon its initial installation. The instructions then tell the installer to change the combo - which many people do not do.

Another consideration for your TLP craft, is that if there are multiples of these locks in the same facility, or installed by the same person/company etc. they usually all use the same code. And even if the code changes, it will often not change drasticlly. Multiple secure facilities that I have had access to during code changes have done things like changing the codes from: (12)(34)5 -to- (13)(24)5 -to (24)(13)5 etc. I've also frequented facilities that have changed their Kaba codes from 1234, to 1423, to 1324, to 4213, etc.

Do not use this information as an excuse to get tunnel vision during an operation! Almost every locked obstacle has multiple tools/techniques and methods that can gain you access. Don't just jump right to trying a long list of codes when a simple latch slip would be quicker. Remember to keep your operational objectives in mind and to choose a tool/technique pair that is likely to be effective, efficient, and reasonable in relation to your operation.

And just because you know of an easy or surefire way to open up a specific lock, does not mean that you *should* open up that lock. You are and should always be accountable for your actions. Make sure that you are using this book and these skills in a morally accountable manner.

Try **(24)3** first.

1	(15)	251	513	(24)5	1(24)
2	(23)	253	514	(25)1	3(24)
3	(24)	254	521	(25)3	5(24)
4	(25)	312	523	(25)4	1(25)
5	(34)	314	524	(34)1	3(25)
12	(35)	315	531	(34)2	4(25)
13	(45)	321	532	(34)5	1(34)
14	123	324	534	(35)1	2(34)
15	124	325	541	(35)2	5(34)
21	125	341	542	(35)4	1(35)
23	132	342	543	(45)1	2(35)
24	134	345	(12)3	(45)2	4(35)
25	135	351	(12)4	(45)3	1(45)
31	142	352	(12)5	3(12)	2(45)
32	143	354	(13)2	4(12)	3(45)
34	145	412	(13)4	5(12)	(123)
35	152	413	(13)5	2(13)	(124)
41	153	415	(14)2	4(13)	(125)
42	154	421	(14)3	5(13)	(134)
43	213	423	(14)5	2(14)	(135)
45	214	425	(15)2	3(14)	(145)
51	215	431	(15)3	5(14)	(234)
52	231	432	(15)4	2(15)	(235)
53	234	435	(23)1	3(15)	(245)
54	235	451	(23)4	4(15)	(345)
(12)	241	452	(23)5	1(23)	1234
(13)	243	453	(24)1	4(23)	1235
(14)	245	512	**(24)3**	5(23)	1243

1245	2315	3254	4325	5413	(15)43
1253	2341	3412	4351	5421	(23)14
1254	2345	3415	4352	5423	(23)15
1324	2351	3421	4512	5431	(23)41
1325	2354	3425	4513	5432	(23)45
1342	2413	3451	4521	(12)34	(23)51
1345	2415	3452	4523	(12)35	(23)54
1352	2431	3512	4531	(12)43	(24)13
1354	2435	3514	4532	(12)45	(24)15
1423	2451	3521	5123	(12)53	(24)31
1425	2453	3524	5124	(12)54	(24)35
1432	2513	3541	5132	(13)24	(24)51
1435	2514	3542	5134	(13)25	(24)53
1452	2531	4123	5142	(13)42	(25)13
1453	2534	4125	5143	(13)45	(25)14
1523	2541	4132	5213	(13)52	(25)31
1524	2543	4135	5214	(13)54	(25)34
1532	3124	4152	5231	(14)23	(25)41
1534	3125	4153	5234	(14)25	(25)43
1542	3142	4213	5241	(14)32	(34)12
1543	3145	4215	5243	(14)35	(34)15
2134	3152	4231	5312	(14)52	(34)21
2135	3154	4235	5314	(14)53	(34)25
2143	3214	4251	5321	(15)23	(34)51
2145	3215	4253	5324	(15)24	(34)52
2153	3241	4312	5341	(15)32	(35)12
2154	3245	4315	5342	(15)34	(35)14
2314	3251	4321	5412	(15)42	(35)21

(35)24	2(15)4	5(34)2	35(14)	15(34)	(15)(34)
(35)41	3(15)2	1(35)2	52(14)	21(34)	(23)(14)
(35)42	3(15)4	1(35)4	53(14)	25(34)	(23)(15)
(45)12	4(15)2	2(35)1	23(15)	51(34)	(23)(45)
(45)13	4(15)3	2(35)4	24(15)	52(34)	(24)(13)
(45)21	1(23)4	4(35)1	32(15)	12(35)	(24)(15)
(45)23	1(23)5	4(35)2	34(15)	14(35)	(24)(35)
(45)31	4(23)1	1(45)2	42(15)	21(35)	(25)(13)
(45)32	4(23)5	1(45)3	43(15)	24(35)	(25)(14)
3(12)4	5(23)1	2(45)1	14(23)	41(35)	(25)(34)
3(12)5	5(23)4	2(45)3	15(23)	42(35)	(34)(12)
4(12)3	1(24)3	3(45)1	41(23)	12(45)	(34)(15)
4(12)5	1(24)5	3(45)2	45(23)	13(45)	(34)(25)
5(12)3	3(24)1	34(12)	51(23)	21(45)	(35)(12)
5(12)4	3(24)5	35(12)	54(23)	23(45)	(35)(14)
2(13)4	5(24)1	43(12)	13(24)	31(45)	(35)(24)
2(13)5	5(24)3	45(12)	15(24)	32(45)	(45)(12)
4(13)2	1(25)3	53(12)	31(24)	(12)(34)	(45)(13)
4(13)5	1(25)4	54(12)	35(24)	(12)(35)	(45)(23)
5(13)2	3(25)1	24(13)	51(24)	(12)(45)	(123)4
5(13)4	3(25)4	25(13)	53(24)	(13)(24)	(123)5
2(14)3	4(25)1	42(13)	13(25)	(13)(25)	(124)3
2(14)5	4(25)3	45(13)	14(25)	(13)(45)	(124)5
3(14)2	1(34)2	52(13)	31(25)	(14)(23)	(125)3
3(14)5	1(34)5	54(13)	34(25)	(14)(25)	(125)4
5(14)2	2(34)1	23(14)	41(25)	(14)(35)	(134)2
5(14)3	2(34)5	25(14)	43(25)	(15)(23)	(134)5
2(15)3	5(34)1	32(14)	12(34)	(15)(24)	(135)2

(135)4	3(245)	15342	31245	41523	52413
(145)2	1(345)	15324	31254	41532	52431
(145)3	2(345)	15423	31452	42315	52134
(234)1	(1234)	15432	31425	42351	52143
(234)5	(1235)	21345	31524	42513	53412
(235)1	(1245)	21354	31542	42531	53421
(235)4	(1345)	21453	32451	42135	53124
(245)1	(2345)	21435	32415	42153	53142
(245)3	12345	21534	32514	43512	53214
(345)1	12354	21543	32541	43521	53241
(345)2	12453	23451	32145	43125	54123
4(123)	12435	23415	32154	43152	54132
5(123)	12534	23514	34512	43215	54213
3(124)	12543	23541	34521	43251	54231
5(124)	13452	23145	34125	45123	54312
3(125)	13425	23154	34152	45132	54321
4(125)	13524	24513	34251	45213	(12)345
2(134)	13542	24531	34215	45231	(12)354
5(134)	13245	24135	35124	45312	(12)435
2(135)	13254	24153	35142	45321	(12)453
4(135)	14523	24351	35241	51234	(12)534
2(145)	14532	24315	35214	51243	(12)543
3(145)	14235	25134	35412	51324	(13)245
1(234)	14253	25143	35421	51342	(13)254
5(234)	14352	25341	41235	51423	(13)425
1(235)	14325	25314	41253	51432	(13)452
4(235)	15234	25413	41352	52314	(13)524
1(245)	15243	25431	41325	52341	(13)542

(14)235	(25)413	4(13)25	1(25)34	53(12)4	31(24)5
(14)253	(25)431	4(13)52	1(25)43	54(12)3	35(24)1
(14)325	(34)125	5(13)24	3(25)14	24(13)5	51(24)3
(14)352	(34)152	5(13)42	3(25)41	25(13)4	53(24)1
(14)523	(34)215	2(14)35	4(25)13	42(13)5	13(25)4
(14)532	(34)251	2(14)53	4(25)31	45(13)2	14(25)3
(15)234	(34)512	3(14)25	1(34)25	52(13)4	31(25)4
(15)243	(34)521	3(14)52	1(34)52	54(13)2	34(25)1
(15)324	(35)124	5(14)23	2(34)15	23(14)5	41(25)3
(15)342	(35)142	5(14)32	2(34)51	25(14)3	43(25)1
(15)423	(35)214	2(15)34	5(34)12	32(14)5	12(34)5
(15)432	(35)241	2(15)43	5(34)21	35(14)2	15(34)2
(23)145	(35)412	3(15)24	1(35)24	52(14)3	21(34)5
(23)154	(35)421	3(15)42	1(35)42	53(14)2	25(34)1
(23)415	(45)123	4(15)23	2(35)14	23(15)4	51(34)2
(23)451	(45)132	4(15)32	2(35)41	24(15)3	52(34)1
(23)514	(45)213	1(23)45	4(35)12	32(15)4	12(35)4
(23)541	(45)231	1(23)54	4(35)21	34(15)2	14(35)2
(24)135	(45)312	4(23)15	1(45)23	42(15)3	21(35)4
(24)153	(45)321	4(23)51	1(45)32	43(15)2	24(35)1
(24)315	3(12)45	5(23)14	2(45)13	14(23)5	41(35)2
(24)351	3(12)54	5(23)41	2(45)31	15(23)4	42(35)1
(24)513	4(12)35	1(24)35	3(45)12	41(23)5	13(45)2
(24)531	4(12)53	1(24)53	3(45)21	45(23)1	12(45)3
(25)134	5(12)34	3(24)15	34(12)5	51(23)4	21(45)3
(25)143	5(12)43	3(24)51	34(12)4	54(23)1	23(45)1
(25)314	2(13)45	5(24)13	43(12)5	13(24)5	31(45)2
(25)341	2(13)54	5(24)31	45(12)3	15(24)3	32(45)1

345(12)	514(23)	213(45)	(35)(12)4	(34)2(15)	4(25)(13)
354(12)	541(23)	231(45)	(35)(14)2	(34)1(25)	1(34)(25)
435(12)	135(24)	312(45)	(35)(24)1	(35)4(12)	2(34)(15)
453(12)	153(24)	321(45)	(45)(12)3	(35)2(14)	5(34)(12)
534(12)	315(24)	(12)(34)5	(45)(13)2	(35)1(24)	1(35)(24)
543(12)	351(24)	(12)(35)4	(45)(23)1	(45)3(12)	2(35)(14)
245(13)	513(24)	(12)(45)3	(12)5(34)	(45)2(13)	4(35)(12)
254(13)	531(24)	(13)(24)5	(12)4(35)	(45)1(23)	1(45)(23)
425(13)	134(25)	(13)(25)4	(12)3(45)	3(12)(45)	2(45)(13)
452(13)	143(25)	(13)(45)2	(13)5(24)	4(12)(35)	3(45)(12)
524(13)	314(25)	(14)(23)5	(13)4(25)	5(12)(34)	(123)45
542(13)	341(25)	(14)(25)3	(13)2(45)	2(13)(45)	(123)54
235(14)	413(25)	(14)(35)2	(14)5(23)	4(13)(25)	(124)35
253(14)	431(25)	(15)(23)4	(14)3(25)	5(13)(24)	(124)53
325(14)	125(34)	(15)(24)3	(14)2(35)	2(14)(35)	(125)34
352(14)	152(34)	(15)(34)2	(15)4(23)	3(14)(25)	(125)43
523(14)	215(34)	(23)(14)5	(15)3(24)	5(14)(23)	(134)25
532(14)	251(34)	(23)(15)4	(15)2(34)	2(15)(34)	(134)52
234(15)	512(34)	(23)(45)1	(23)5(14)	3(15)(24)	(135)24
243(15)	521(34)	(24)(13)5	(23)4(15)	4(15)(23)	(135)42
324(15)	124(35)	(24)(15)3	(23)1(45)	1(23)(45)	(145)23
342(15)	142(35)	(24)(35)1	(24)5(13)	4(23)(15)	(145)32
423(15)	214(35)	(25)(13)4	(24)3(15)	5(23)(14)	(234)15
432(15)	241(35)	(25)(14)3	(24)1(35)	1(24)(35)	(234)51
145(23)	412(35)	(25)(34)1	(25)4(13)	3(24)(15)	(235)14
154(23)	421(35)	(34)(12)5	(25)3(14)	5(24)(13)	(235)41
415(23)	123(45)	(34)(15)2	(25)1(34)	1(25)(34)	(245)13
451(23)	132(45)	(34)(25)1	(34)5(12)	3(25)(14)	(245)31

(345)12	3(145)2	34(125)	31(245)	(45)(123)	(1345)2
(345)21	1(234)5	43(125)	12(345)	(35)(124)	(2345)1
4(123)5	5(234)1	25(134)	21(345)	(34)(125)	5(1234)
5(123)4	1(235)4	52(134)	(123)(45)	(25)(134)	4(1235)
3(124)5	4(235)1	24(135)	(124)(35)	(24)(135)	3(1245)
5(124)3	1(245)3	42(135)	(125)(34)	(23)(145)	2(1345)
3(125)4	3(245)1	23(145)	(134)(25)	(15)(234)	1(2345)
4(125)3	1(345)2	32(145)	(135)(24)	(14)(235)	(12345)
2(134)5	2(345)1	15(234)	(145)(23)	(13)(245)	-END-
5(134)2	45(123)	51(234)	(234)(15)	(12)(345)	
2(135)4	54(123)	14(235)	(235)(14)	(1234)5	
4(135)2	35(124)	41(235)	(245)(13)	(1235)4	
2(145)3	53(124)	13(245)	(345)(12)	(1245)3	

KWIKSET SMARTKEY LOCKS

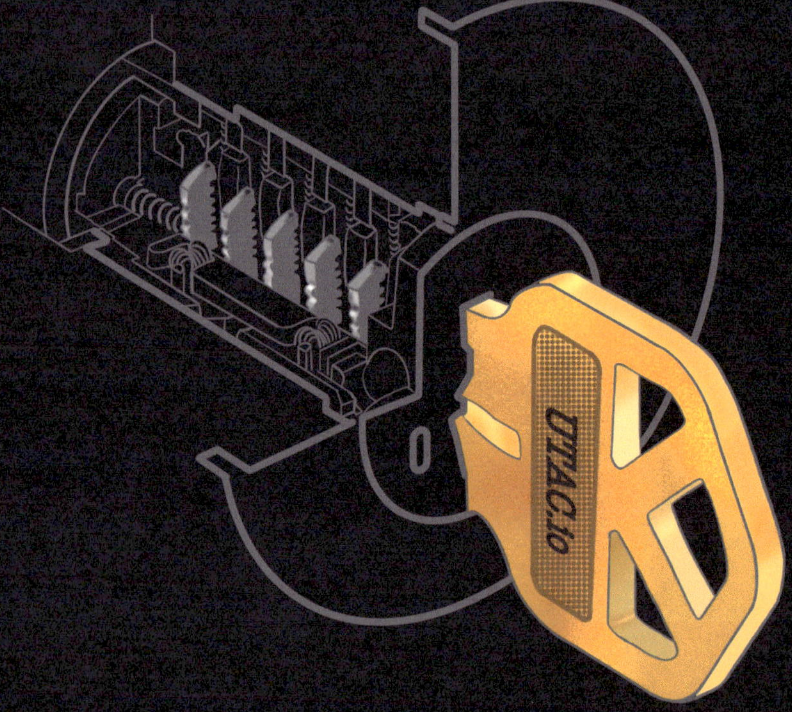

Kwikset "SmartKey" type locks are unique. They are designed to be a residential consumer grade lock featuring a "set your own combo" function, allowing users to change the bitting of the lock (rekey the lock) without a locksmith. The process can be done in about 60 seconds by even an untrained user and all it takes is the old key, a combo change tool (costs about $1) and the new key that you want your lock to be set to.

These are quickly becomming the most popular and rapidly growing type of residential lock in the U.S. market. These are particularly common with new building construction where construction companies can very easily change the locks without any added cost or without any locksmiths or confusing key tracking information.

5-10 years ago you could still be pretty successful with non-destructive entry even without having any tools or techniques in your toolbox for these SmartKey locks. With the growing popularity of these locks today, you are going to fail pretty quickly and often on your journey if you are not prepared to handle these locks.

There are multiple generations of this lock on the market, each one has slightly different tools and techniques that it is susceptable to for NDE.

<u>Some of the most important things to know right up front are</u>:

•SmartKey locks are **not** pin-tumbler locking mechanisms, and they do not function the same as pin-tumblers (although they look similar on the outside, and they use a pin-tumbler type key).

•SmartKey locks can **not** be picked using the same *basic* pin-tumbler picking and raking methods.

•They **can** be picked, but it requires special tools and special techniques that are not the same as standard pin-tumbler techniques.

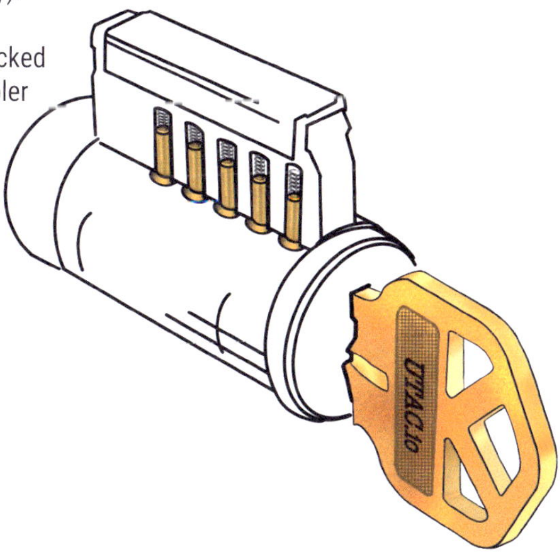

SMARTKEY CAMERA DECODER (& KEY GENERATION)

(Position 1) (Position 2) (Position 3) (Position 4) (Position 5)

*Example: Photos showing the wafer heights within each position of this lock.

For most of the locked obstacles you will encounter in the field, there are multiple non-destructive entry tools & techniques that you can leverage to gain entry. These Kwikset SmartKey type locks are no exception. Smartkey locks are identifiable by the small vertical rectangular hole next to the keyway.

Before picking or decoding a lock, I will almost always first seek an open door or an open window, I'll look for a hidden key stored nearby, and I'll check to see if there is a known bypass method for the lock that I've encountered. If none of those apply, then I'll prepare to apply a Kwikset SmartKey specific entry.

If I understand basic KeyGen principles, I'll grab my KeyGen tools and SmartKey camera. It shows you the lock's bitting information; which position of the lock you're looking at and what the height of the key needs to be at that position.

Almost as simple as "Match the image on the screen to your reference card." You will very quickly learn, though, that entries in the field are very different than social media. First, the camera needs to be charged -- and what do you do on scene if your camera is dead, or dies during your entry? Then, you need to establish a solid connection between your phone and your camera, sometimes obscure phone settings can make this an arduous process to resolve. Then, it actually takes quite a bit of practice to understand the images you are seeing on your phone screen because they aren't always 100% clear.

Also, if you decode the lock correctly the first time, but don't know how to generate a key that is cut to code, then what will you do with that lock's bitting information? We published an entire book on how to generate keys using field expedient methods, but even if you don't have the tools or the training, you can still bring a bitting code to a local locksmith who can easily cut you a key based off of a simple string of numbers.

Remember, entry speed is always relative to your unique situation. If your only other solid option for entry will take over an hour to arrive or to apply to the situation, then using a camera and cutting a key on scene in 3 minutes is *lightning fast*!

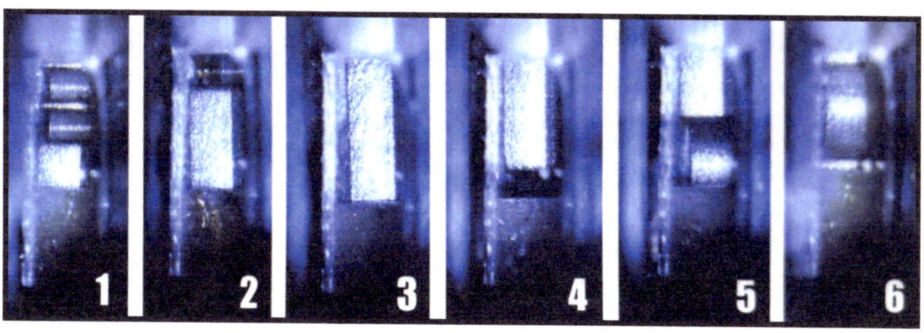

KEY GENERATION

If you are able to use your camera decoder to pull the bitting (the key/lock measurements) for a SmartKey lock, do you have a plan for what to do next? We have an entire book volume that provides pathways for how to generate keys in the field, but for a quick reference, we can give you three simple solution avenues that you may want to consider for these SmartKey locks.

<u>Outsource</u>: Once you have a confident guess at the bitting code for the SmartKey lock that you are up against, you can either go to a locksmith, or have a locksmith come to you to "cut a key to code" using their locksmith grade tools. This **might be fast** if you have planned ahead, or if a locksmith business is very close to your operation, but this **might also be super-slow** based upon several factors. The upside is that beyond visually decoding the lock, the specialized craft of key generation is done by a professional and **you don't need any infrastructure or knowledge** of how to generate a key.

<u>Light and Flexible</u>: Using **small** handheld tools (photo opposite: right side of the photo) that are relatively **inexpensive**, you can "clip" a key with something as minimal as a set of vice grip pliers, a Spacing and Depth key set (keys cut to a known depth), a small hand file, and some key blanks. You can get faster with this as you progress, but it **takes a while** to become able to apply this method reliably and repeatably.

<u>Dedicated KeyGen</u>: Using an **expensive**, **heavy**, **bulky** system (photo opposite: left side of photo), you can cut keys to perfect factory specifications very **rapidly**, very **accurately**, and very **reliably** using your own locksmith-grade key cutting tools. This takes a dedicated budget, dedicated training and knowledge, and maintenance, etc. These types of expensive key originating tools are not for everyone, but maybe they are right for you, depending on your operational expectations and infrastructure.

You get to decide if you want to learn any KeyGen methods at all, and if you do, you get to decide how deep you will go. There's nothing wrong with just learning a few Tactical Lock Picking techniques and owning a few tools, but you can take this skill set to incredible depths and you can increase your capabilities to top-tier levels if you choose to. Either way, we are glad that you are here to learn.

If you decide to add key making to your operations (which has a ton of benefits) consider checking out our book *Tactical Lock Picking Vol. 2 - Field Key Generation*. There is a lot of myth and misinformation surrounding lock picking (especially lock picking during emergencies and operations) and there is just as much myth surrounding key generation. Our goal is to prepare people to use all types of NDE methods *in the field*.

UTAC'S KWIKSET
FIELD TRYOUT KEYS

Since your basic tension wrench and rake method, and simple SPP method, is unlikely to work on Kwikset Smartkey locks, our Kwikset Smartkey Field Tryout Key Set might be a very useful solution to fold into your application toolbox. This set leverages the half-position bittings that the SmartKey lock allows for wiggle-room. This set does not work on the Kwikset pin-tumbler locks.

At the simplest level, you clip both bags to your belt or belt loops, one bag will have all of the **243 keys**, the other will be empty. You pick up one key from the "fresh" pouch, try it in the keyway with some jiggling and twisting, and if it doesn't work you discard it into the "used" pouch, until you find the correct key. Upon finding the correct key, if you require repeat or future access, you can clip your working key into the red "key leash" that is clipped into your bags so that you can easily isolate the correct key for your target lock without it getting lost in the mix of non-working keys.

At the advanced level, you can access a .pdf file that is searchable that can be leveraged to help you find *exactly* which key should work in the lock if you have pulled bitting with a camera (or other method). And the .pdf can help you find out which exact bittings your half-cut key represents.

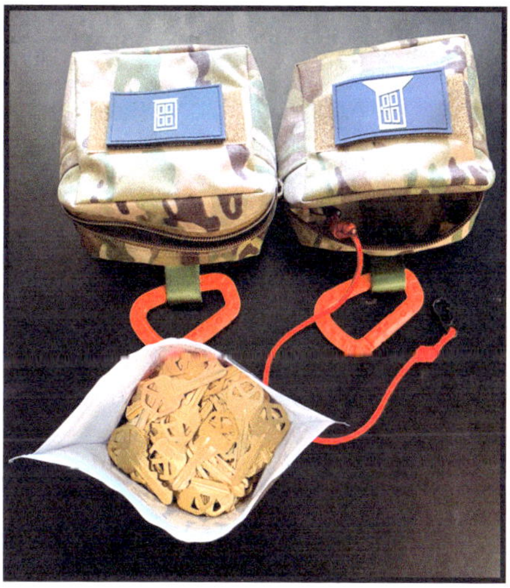

This set was not designed to be an everyday carry, or carried all the time. This was designed to be grabbed from a closet or a vehicle trunk when needed for an entry; or clipped onto your second line gear bag (2LG) if you expect to be dealing with Kwikset SmartKey locks frequently.

Your Tool&Technique pairing choice should also fully consider your operational or emergency framing. If destructive entry is not your best option, and if you do have the time to try these keys in a keyway, then this set may be incredibly useful to you.

This kit is available at UTAC.io with multicam or black bags.
Comes with two glow in the dark UTAC door velcro pvc patches, two plastic carabiners, a key leash and 243 tryout keys.
These keys and pouches are cut and sewn **in central Georgia, U.S.A.**
This QR code, right, will take you directly to our product page.

WARDED LOCKS

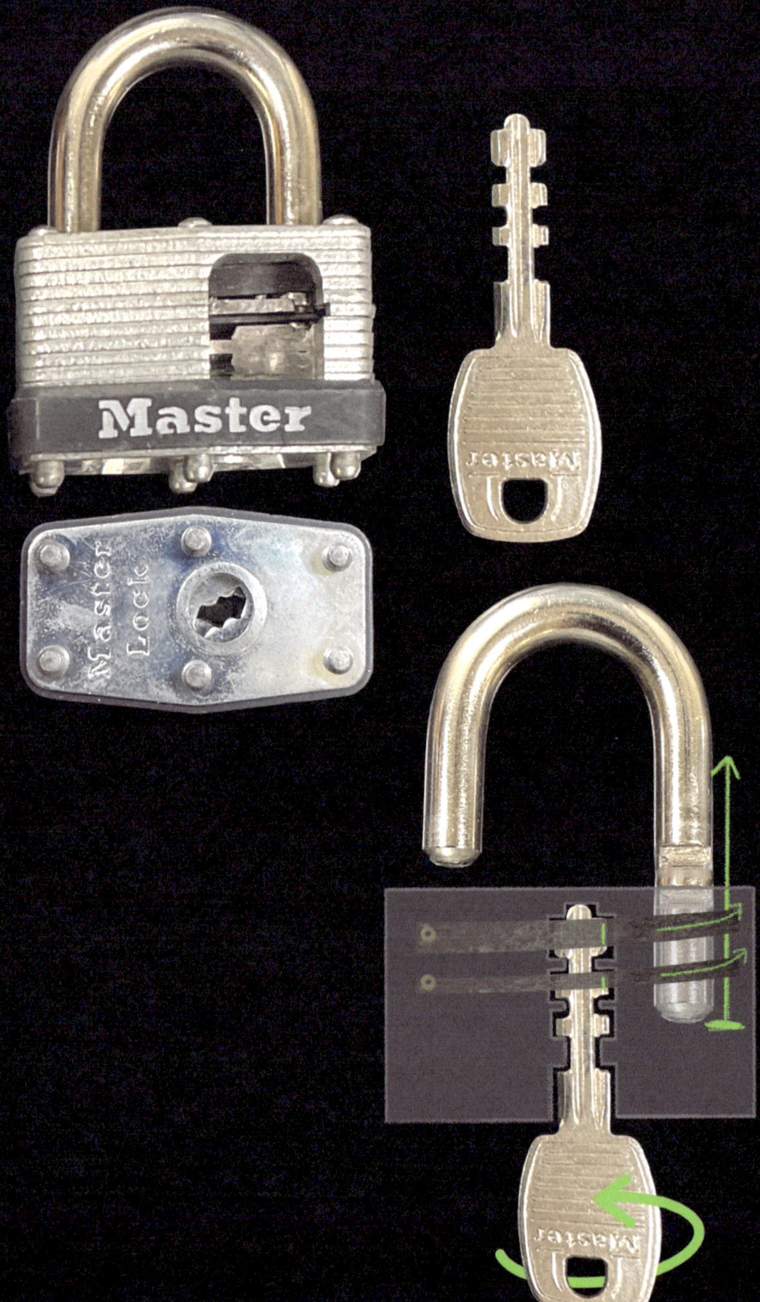

Warded locks are one of the oldest locking types known to history. Depending on how you count, there's really only one moving part, maybe two. Most modern warded padlocks consist of several chambers (or "wards") cut into the keyway so that only the right key can fit into the wards and then turn, which lifts one or sometimes two flat metal springs, allowing the shackle to pop open.

Warded picks are so easy to use and are so effective at opening up these low-cost and low-security padlocks that you don't even need a "60 second crash course," just a 10 second crash course will do.

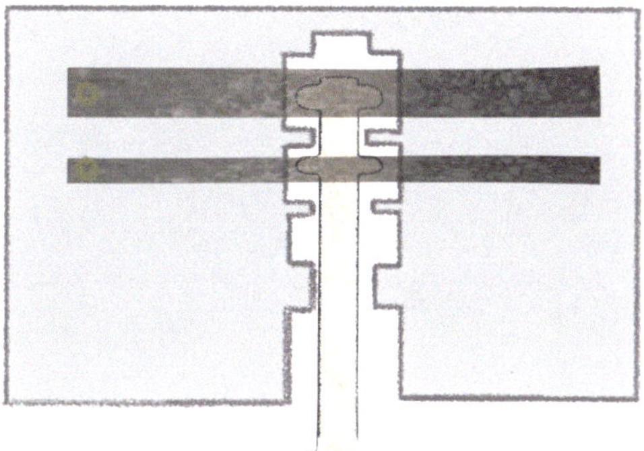

Pick one of the warded pick profiles, insert into the keyway, and twist it around randomly for a few seconds. If it doesn't work, switch to the next pick profile and repeat.

Warded padlocks are not incrediby common, but they are incrediby inexpensive, often the least expensive padlocks on the shelf at any of your local hardware stores. Because they are inexpensive, they are *common enough* that it is worth folding some warded picks into your 1st Line Gear, or at least your 2nd Line Gear bag.

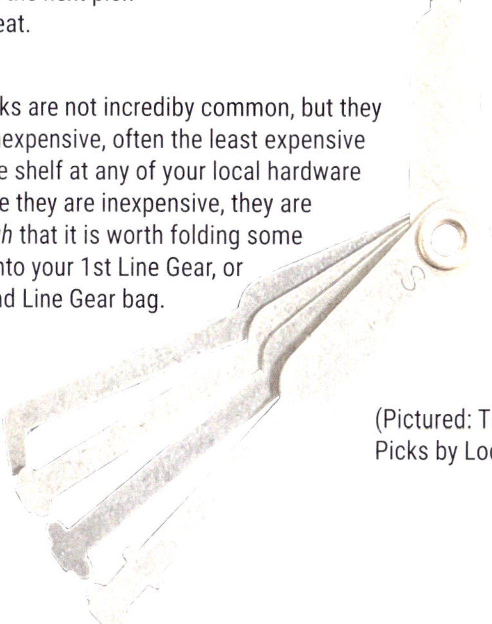

(Pictured: The 8/60 Warded Picks by LockPickTools.com)

*This is one of the very few padlock tool/technique combos where it actually *does* help your entry chances if you pull on the shackle a little bit while executing this technique.

WAFER LOCKS

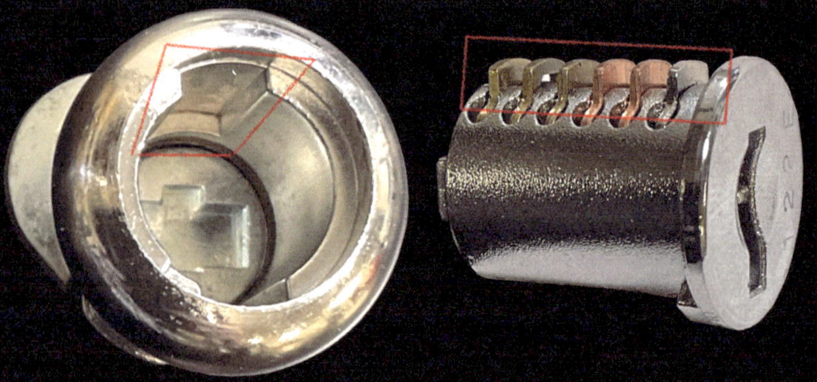

Wafer locks function similarly to pin tumbler locks, except it's not pin stacks that need to be lifted to the right height in order to clear obstructions from the shear line allowing the plug to spin. Wafer locks have wafers that protrude up and out of the plug that need to be raised/lowered to the correct position in order to clear the shear line. Not all, but some of the basic raking tools and techniques for pin-tumbler locks will also work on some wafer-type locks.

Identifiable by what they're mounted to - storage containers, file cabinets, inside push-button handles, on machinery panels (seen below).

Also identifiable by looking into the keyway. Pin tumbler locks will have one rounded pin extended into the top of the keyway; wafer locks will have a flat chunk of metal on one or both sides of the keyway, top and/or bottom (seen below).

Pin Tumbler

Wafer

Wafer locks are *usually* very inexpesive and very low security. The basic concept for opening is the same as the raking technique we showed in the beginning of this manual. Apply some light twisting tension on the keyway, and while holding that tension, use a rake-type tool to randomly raise and lower the *wafers* (in this instance) until you achieve an unlocking.

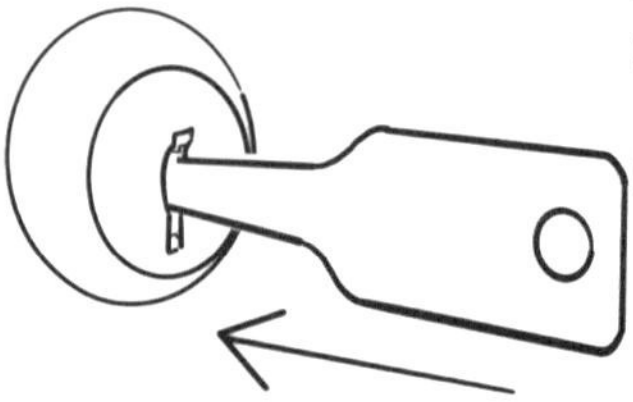

To use a "jiggler" technique on a low security wafer lock, only a quick how-to crash course is required.

First, you insert any jiggler profile of your choice.

Then, you apply some light twisting pressure, and hold that pressure.

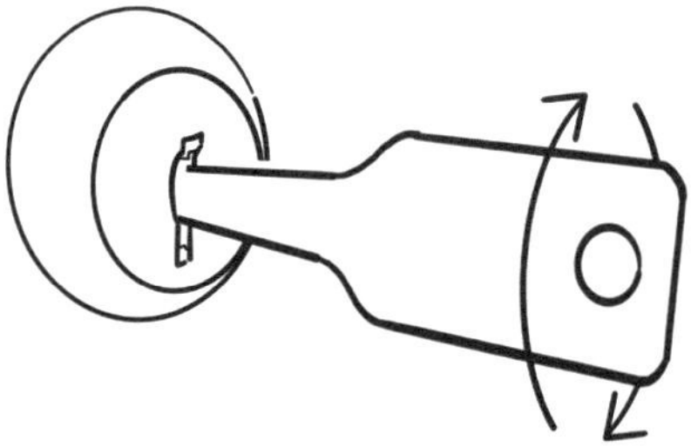

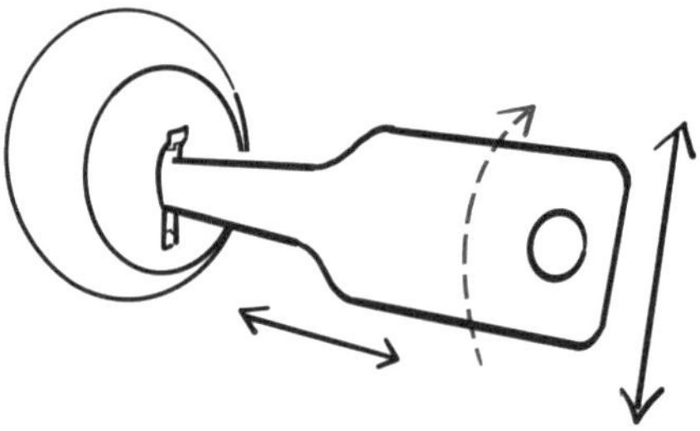

Now, while keeping that light twisting pressure, slide the jiggler in and out, and rock it up and down; ie "jiggle" it.

The good news is that an overwhelming majority of the actual wafer locks that you are going to come across are low security, or medium-level security at best. And there is some overlap too, while these jigglers are a preferred tool/technique combo for most wafer locks, your standard raking method with a tension wrench and a Bogota® style rake will often grant you an entry as well.

On most days, even as a beginner or medium level Tactical Lock Picking operator, most of these inexpensive and common wafer-style locks will just melt open for you with little to no stress. But there are many high-level TLP practitioners that can just get stumped on the simplist of wafer locks even with a full compliment of picking/raking and jiggler tools. It happens.

Wafer locks are common enough, and openable enough, that your 2nd Line Gear bag would be naked without a few jiggler tools in it.

JIGGLER TOOL EXAMPLES

A pretty standard "Jiggler Set." Many manufacturers make this set with identical or very similar tool shapes and sizes.

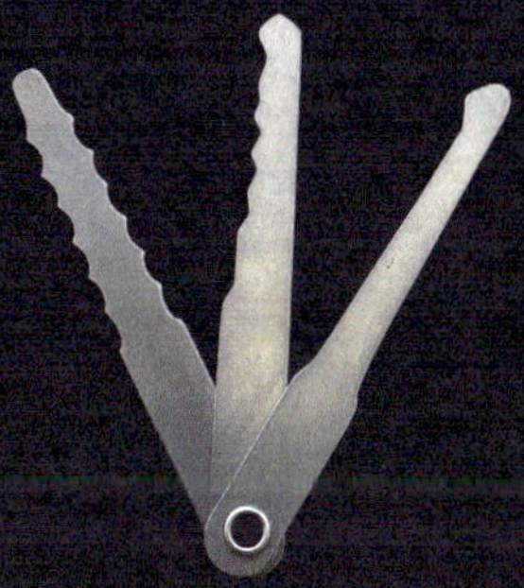

This tiny set is included with our "Covert 8/60" pocket sets and also has a QuickStick as a fourth tool that can swivel out.

"Covert Jigglers" are easy to carry on a key ring for EDC access.

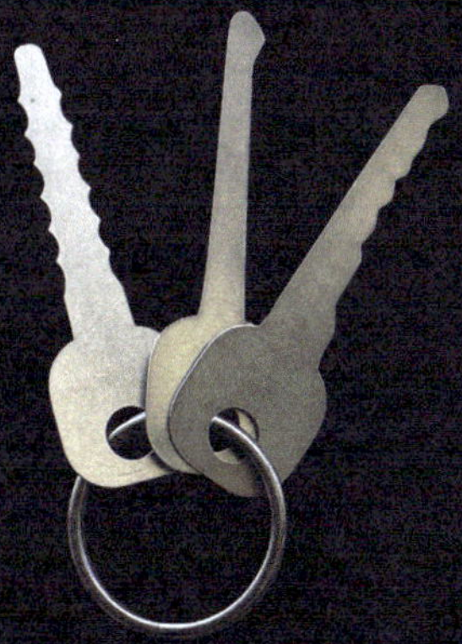

*All three of these sets are available at LockPickTools.com

KEYED ALIKE

"Keyed Alike" keys are a product of bureaucratic functions. Many companies that make things like RV's, toolboxes, file cabinets etc. and don't sit down at the drawing board to create their product and then also open up a file for "fresh, new lock design." They instead just buy a bunch of identical locks that go into one bucket, and a bunch of identical keys that go into another bucket, and when their product is being assembled, they just pick up a lock from one box and a key from the next box and they slam them into their product and they move on. This results in all of their products off of the assembly line being able to be opened with the same identical key.

As a Tactical Lock Picking entry specialist, you can design your own "Keyed Alike" key set that is specific to your Area of Operations (AO). Take into account which types of locked obstacles you are likely to enconter, as well as what your personal capability is with all of the different sub-categories of entry studies. If you know nothing of electronics, don't try to plug a wire loop connecter into a computer panel for a vehicle gate. If you work with heavy commercial machinery, maybe you should purchase the known keys for those brands/makes/models of machinery.

But, like most of TLP, this is not "plug and play." You'll still have to do your homework. Homework is required to setup your set and to test it and make sure it meets your expectations before relying on it on operations or during emergencies; then make sure that you're staying with the curve as far as industry changes are concerned. As companies discover the flaws in this keyed alike circumstance, they may change to a new key, or multiple limited keys for their locks. Also, technology will change and companies will use different locking mechanisms, and they'll also have changes in availability and prices for their lock and key sources.

Consistent with the rest of our TLP curriculum, your keyed alike key set is also a percentage game. Is it likely that this tool will work in the field? Did you test it with as many variables as possible before relying upon it in the field? Are you consistently tracking data for when you get to use this specific tool, how frequenetly you use it, and when it works; and more importantly, when it *doesn't*?

The most popular two keys that can get you started are the **CH751** and the **C415A**.The CH751 (pictured opposite) is one of the most common keys for what we call "furniture locks" as in a lock that is mounted to a piece of furniture like a toolbox, or an RV cabinet, or a file cabinet. And the C415A is one of the most common cabinet keys in the U.S., commonly used for cabinets and lock boxes of many different shapes and sizes.

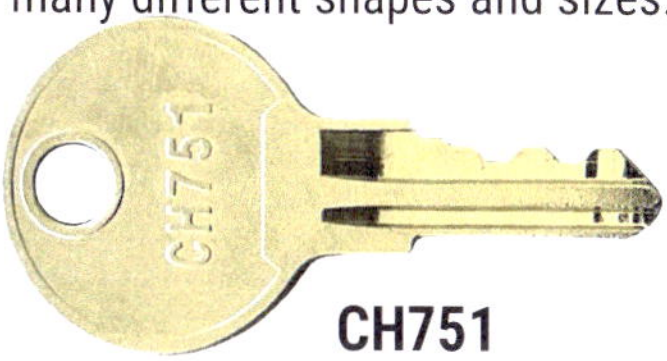

CH751

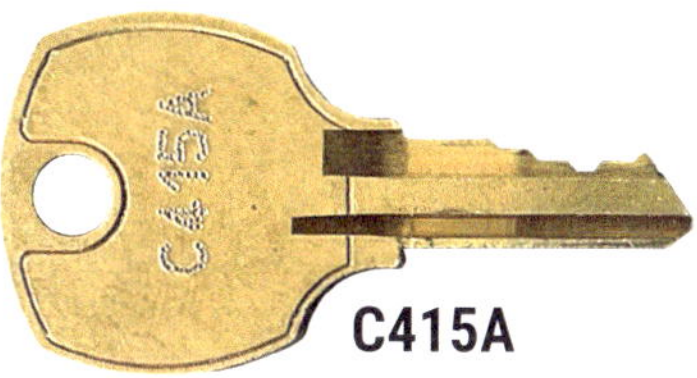

C415A

VEHICLE ENTRIES

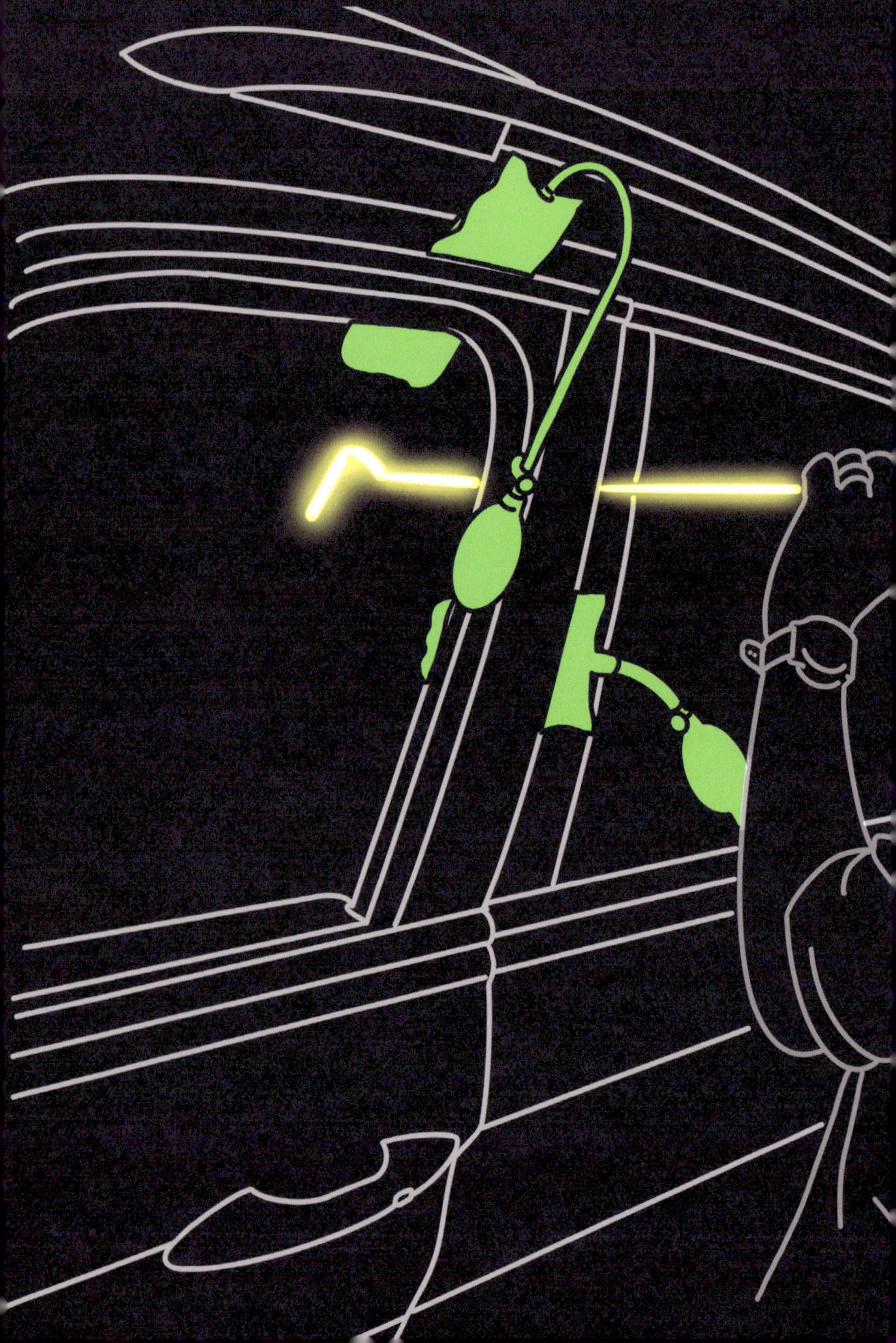

To absolutely maximize your non-destructive vehicle entry capabilities, in a minimalist manner, we recommend our classic vehicle entry loadout. We carry, in our UTAC Vehicle Reach Tool Bag, one semi-rigid reach tool for around $30 (we prefer the "Gold Finger Tool"). And we carry two separate small rigid air-wedge tools of any brand for around $10 each. A three piece kit that will get you into a huge percentage of make/model/year of vehicles.

For entry, get at least one air-wedge partially inserted near the top corner of the driver side or front passenger door; inflate. Then, with a small gap created now, insert your second air-wedge; inflate. With a solid gap created between the door and the vehicle frame, adjust both bags to give yourself a nice gap near the top corner of the door. Then, depending on the make/model/year of vehicle, and the current variables (is the battery dead, are the windows and locks electronic or manual, etc.), insert your reach tool and use it to push, pull, prod etc. to help you gain access.

There are always many variabels to consider with vehicle entries. But with experience, and with many failures, you will start to sharpen your edge in order to make smarter decisions to set yourself up for the most effective and efficient vehicle entries.

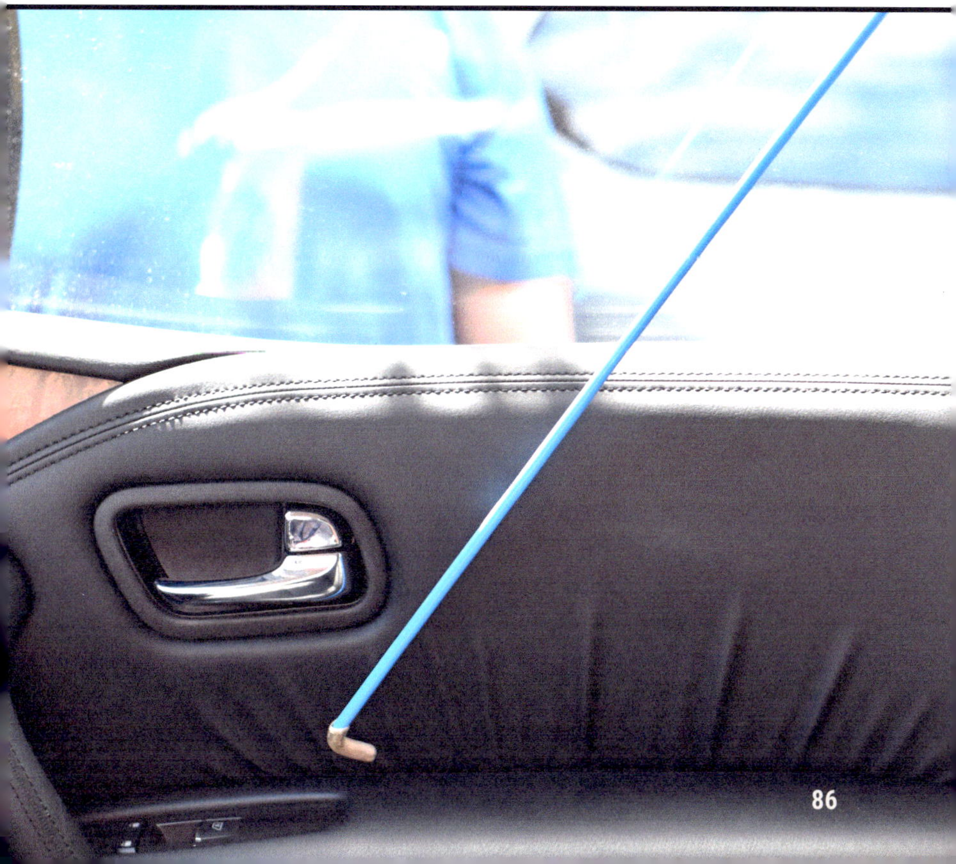

TARGET ASSESSMENT

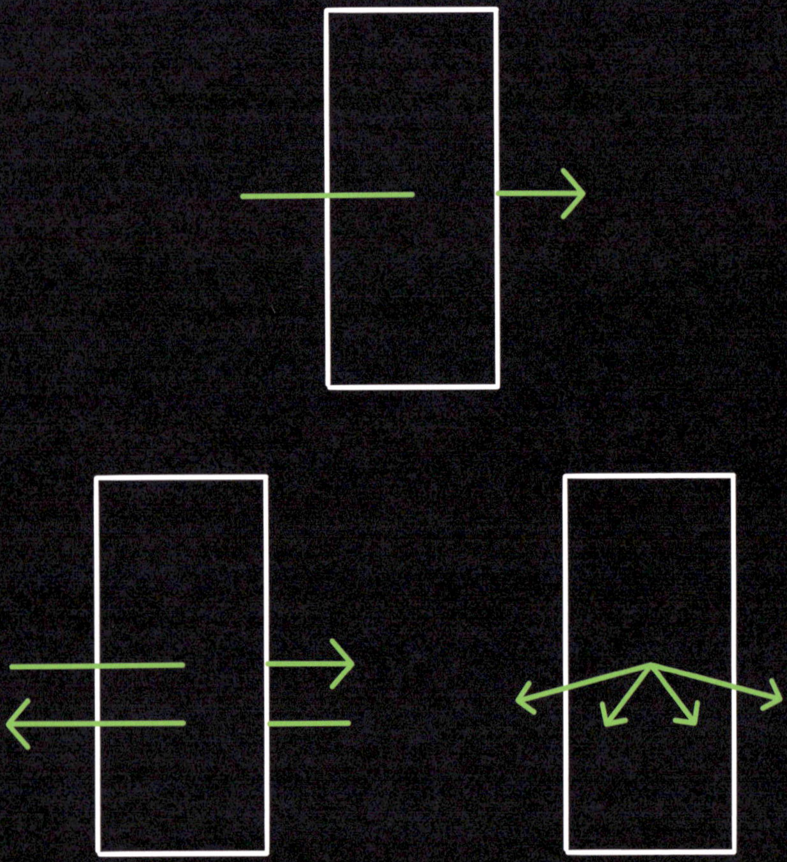

Having a **strong and sound foundational approach** to your skill is crucial for a practicing **field-expert**. The environments in which you expect to be be applying your skills matters; it matters a lot. For the hobbyists that publish their lock picking how-to videos on social media, they have absolutely no requirement to take into account any field environmental issues. To be clear, being a hobbyist is fantastic, and we support those who practice purely as a hobby.

But the fact remains that someone at a desk, in a swivel chair, behind a camera, can drop their picks on the floor 100 times during their filming takes with no real consequense; the field practitioner who is balancing on a pipe to unlock a potential foot path for rescue personnel where kinetic/destructive entry is not feasable, only has to drop a single tool a single time in order for the tool to become forever irretrieveable by the sweeping water below. This means that the "foundational approach" to a social media hobbyist video must take into account camera angles; and the "foundational approach" for the field operator must take into account a stressed timeline, the limited capacity to carry only a few tools to the locked obstacle, the management and packaging and storage and deployment of tools to the obstacle, the slow and deliberate care that must be exercised during the technique application, and even ensuring that they don't get tunnel vision on a single lock and must constantly be assessing what other types of attack vectors may be available like a hidden second keyway somewhere, or a hidden code nearby for combination locks, or certain types of furniture bypasses that may speed up an entry 10x over an expected standard picking attack, and much more!

How can you make the most out of your field entries by keeping a strong Target Assessment foundation in mind?

<u>Gear selection</u>: Which gear should you initially acquire for your craft? And can you carry all of that gear into the field, or should you pack a loadout ahead of time with your most useful tools, or the tools that fit your expected Area of Operations (AO) the best? Also, if you have any intel about the locks you are expected to encounter for a specific operation, would that change your loadout if you had time before your departure to the operation?

<u>Gear deployment</u>: We define this as how you organize your gear within your loadout. Is everything just in one medium sized bag or box? Or do you have specific tools within specific bag(s) and within specific pockets and pouches for specific field-environment reasons? And, once you are at your locked obstacle, how will you be managing the retrieval of tools from your loadout, and how will you be stowing your gear while you are switching from one tool to another (very different from working on a flat desk during a video shoot) and how will you stow your gear after you accomplish an unlock, or after you decide to move from your target, etc.?

Overall Operational Objectives: Are you an agent without operational cover who must appear non-operational by carrying only makeshift entry and escape tools? Are you a first responder who has a wealth of other assets and infrastructure in which to assist you while you are attempting entry? Are you a top tier operator on a CMOE operation where speed and a lack of entry evidence is your primary operational constraint? Or are you merely trying to unlock a pedestrian gate within a fenceline so that you and your team don't have to unnecessarily climb a fence? These all have an impact on the entry decisions that you make (including abandoning NDE and going destructive; or aborting mission alltogether).

Access Objectives: Is the access goal for you, as a single human, to achieve a non-emergency entry into a locked space for a simple administrative reason? Meaning that all you need to do is **gain access**, in no particular rush, for no important reason? "Once I get through this locked obstacle, mission complete!"

Is the access goal for you, as an operational requirement, to traverse in-and-out through this locked obstacle multiple times (either within one sitting, or across many hours, days, weeks etc.)? Meaning that gaining access is not enough; you must **keep access** through this locked obstacle for multiple passings. "Once I get in here I'll need to maintain this unlocked status, since this perimeter door auto-locks on the outside when it closes and I'll be making multiple trips through this access point during this operation."

Or is the access goal for you, as a team member, to **share access** to your locked obstacle with other people? Does your operation require you to create additional keys for other people? Is bypassing a combination lock enough? Or does your operation require you to determine specifically what the combinaion is and then to share that combination with other team members? "Alright team, once you get here the gate will be locked, but I was able to decode the combo which is 2-4-1-3-Enter."

Being familiar with not just how to *get through locks* but also how to ensure that locks *remain unlocked* when required, and how to *share locked access* when necessary can help you level up your skill set as an entry expert. And, as a TLP operator, you can still put a win in the win column if you are tasked with entry and find a nearby open or unlocked window to crawl through.

Tactical Lock Picking does not exist in a vacuum or solely behind a video camera for the social likes. TLP was designed from day one to support real world operations, whether you are an individual unsupported operator, whether you are on a high level operational team. Whether you are undercover or behind enemy lines, or whether you are a county plumber, a soil scientist or a soccer mom who has legitimate access through a locked vehicle gate but happens to find a padlock and a chain on the gate and is merely "locked out" but with entry permission, and no key.

KATY, Texas. (July 2023)

7 Target Assessment Steps

1. Do you have the **Moral Right** to make entry? If yes, continue. If no, stop. And remember, moral and legal are not always the same thing. And, triple check that you have identified the correct target!

2. What is the **Speed** at which entry is required? Is this an administrative entry, is this an operation where faster entry is helpful but not absolutely necessary yet, or is this a "seconds matter" outright emergency?

3. What **Attack Vectors** are known to you at this point? How many attack vectors have you identified so far? Are there others that you can go look for near the target lock? Are there attack vectors that might not be known to you?

4. **Manual Check** locked/unlocked status. For padlocks, door locks, windows, gates, and everything else that you *can* check, you *should* check. "Try the knob, take a deep breath, then try again *harder*!"

5. **Tool Availability**? What tools are on you, near you, can be retrieved by you or even brought to you, or can be makeshifted within your environment.

6. **Now or Later**? Should you start your hands on entry attempt now, or do more recon and then decide which vector you will attack through first?

7. **Other Resources** (phone a friend, internet search, locksmith, etc.); take a big "**Reset breath**;" **Retrace** all your steps so far. Zoom out and take another look at the problem in its entirety. Consider if you've missed something. Consider if you just need more time and should continue with your best tool and technique selection so far, and always be reassesing the mission overview. Timelines change, entry permissions change, tool availability changes, infrastructure and support can change, environmental factors like heat and cold, rain and wind, light and darkness will also play a role in your decision making.

BATON ROUGE, La. (July 2023)

IDENTIFYING EASY KEYWAYS TO PICK (STANDARD PIN TUMBLER LOCKS)

Kwikset Residential Keyways (& Similar)

- Easily identifiable as low-cost.
- Simple “zig-zag-zig” keyway.
- Wide keyway with a lot of room to maneuver tools around.
- Arguably the most common residential keyways in the U.S. (or close to it)

Schlage Basic Keyways (& Similar)

- Easily identifiable as low-cost.
- Simple “zag-zig-zag” keyway.
- Thick keyway with a lot of room to maneuver tools around.
- Slightly more difficult than standard Kwiksets, but not by much.
- Very common on homes and in commercial settings.

Basic Padlock Keyways

- Easily identifiable as low-cost.
- Keyway has multiple zigs, but is still wide open, leaving a lot of room for tools.
- With time and experience, you will be able to quickly identify several very common inexpensive padlocks like this.
- Found in almost every “big box” store.

TLDR: The more common and the more low-cost a lock/keyway is, the easier it is to pick open. There are exceptions to everything, always.

First, you'll need to know how to look at a lock and verify that it is in fact a pin-tumbler type lock. Look for a single pin hanging in the keyway, and that classic zig-zag-zig shape of the keyway. You'll also be able to tell lock difficulty by looking at what the lock is mounted to. If the lock is mounted to a standard cookiecutter type residential home, it's probably a pin tumbler (or a SmartKey lock, which is rapidly growing in popularity).

If the lock is mounted to something low value, like a backyard toolshed of a standard residential home, it's probably low security. If it's a pin tumbler lock mounted to the front door of a high-end jewler store in a fancy part of town, it's probably a much more difficult keyway to pick.

Even if you're not familiar with machinery and fabrication and construction, the more you get different locks into your hands and the more things you try to pick open (successfully or not) the more you'll start to notice the quality of the construction of the lock. You'll notice tighter tolerances, heavier metals, smoother and shinier finishes etc. Usually, not always, the more production value that goes into the machining of the lock, the more difficult the keyway will be to pick open.

If the shape of the pin-tumbler keyway is simple and wide open, that's usually a good sign that it's easy to pick. If the keyway is overly zig-zaggy (we call that a paracentric keyway) then it will be more difficult to get your picks into the keyway (examples on the next page).

Enviornment. Sometimes a worn and weathered and rusted lock will be much harder to pick because everything is sticky and gritty. Other times a lock that has seen a lot of weathering might just fall open as soon as you put your picks into it. Weathering is about a 50/50 shot at making a lock easier or harder to pick, just be aware that it will play some type of role in difficulty.

Positioning is important too. A simple low-security padlock in your lap should be easy to pick. That same low-security padlock that is facing upside down, that you have to pick by reaching through a fence, at night, in the cold rain, will be much harder. Environments always play a factor.

Tools and skills factor in as well. If you are a day-1 practitioner of TLP, every lock in existance may end up being a "difficult lock." As your skill grows, difficult locks will become easier for you. And, for a seasoned practitioner, a lack of tool selection might take an otherwise openable lock and turn it into a struggle.

IDENTIFYING DIFFICULT KEYWAYS (STANDARD PIN TUMBLER LOCKS)

BEST High Security Keyway (& Similar)

- Keyway is very "paracentric" (zig-zaggy across the center line).
- Higher quality machining than the low-cost keyways on the previous pages.
- Found mainly on high security or expensive commercial facilities / gov facilities.
- Still "pickable" but after a long investment in time, training, knowledge, and experience.

Schlage "Everest" (& Similar)

- Keyway is very "paracentric."
- High quality machining.
- Brand, and Everest Logo Stamp, are searchable online, easily identifiable as difficult.
- Commonly found on high security doors.

Yes. If you plan on making the skill of **Tactical Lock Picking** a permanent part of your toolbox, and you plan to use this skill operationally and for emergencies, you should always be practicing to get better and to be able to open more difficult locked obstacles as your skill progresses... Or at least practicing to maintain whatever level you're happy with once you get there.

And, yes. It should also be a part of your skillset to understand the limitation of your skill set, and the limitation of your tools, so that you don't waste your precious time on the scene of an emergency trying to implement a tool & technique combo on something that won't open for you in a million years.

The two locks on the opposite page, the BEST SFIC keyway (Small Format Interchangable Core) and the Schlage Everest, are both "pickable" but in our opinion, there is a zero percent chance that you will open either of these locks with your first $25 Bogota® toolset and your first crash course on the raking technique.

Don't get overwhelmed or discouraged with trying to memorize all of the different makes and models of different types of locks and keyways. Start your research and your practice with the easiest locks to open. Then start looking for those same makes/models in the wild. You'll quickly start to be able to immediately identify locks in the wild that are the same as the ones you've opened before in your practices. Kind of like seeing a lot of white four-door pickup trucks after you buy a white four-door pickup truck.

Remember that all of Tactical Lock Picking is a percentage game. Your ability to start to identify locks on sight will increase with a percentage of accuracy over time. And you'll make mistakes along the way - and THAT'S WHERE YOU'LL LEARN THE MOST!

We're not saying that you shouldn't try to pick something if it looks tough, but we are saying that if you find yourself absolutely brick-walled, then you should try to log that make and model and to dissect your failure and figure out specifically why you were unable to open said lock.

TLDR: You'll quickly start to learn that simple raking and picking will not work on every single lock out there - pay attention when you come across those tough ones and work that into your decision making for future entries.

Anything with "medeco" stamped on it is also well outside of a beginner's skill set.

ADDITIONAL CONSIDERATIONS FOR LOCK ID

Before you set out on a mission to shove your pin-tumbler picking tools into a keyway... make sure that you're dealing with a pin-tumbler lock; and make sure that the keyway is actually a keyway; and that the lock is actually openable with the tool/technique pair that you have selected.

Yes... there is value to trying *something* when you are out of options and short on time and other resources. But, no... you should not be going out and getting *tunnel vision* as soon as you see a hole on the side of a lock; sticking the wrong tool into it without any possibility of an opening.

The Kwikset Smartkey locks are *not* pin-tumbler locks and will not open using your normal, standard, beginner raking and picking methods.

Dimple locks operate in a similar way to how the standard pin-tumbler locks do, but your normal tension wrench and normal Bogota® (or similar) rakes will probably not get you an entry.

The hole on the side of most Master 175's is not a keyway that you can pick open. It is an accessway for a combination change tool, and there is nothing in this accessway that will magically grant you an unlock.

We say this not to discourage you, but rather to ensure you are being pointed in the right direction. Especially as a beginner, the tools and techniques that you learn early on will get you *very* far, but we also want to temper that with the understanding that your beginner tools & techniques are not *magic*, and that they certainly do have limitations.

MAKESHIFT TOOLS

Makeshift Tool Definition Examples

- An item that you find on scene that you can use to help you gain non-destructive entry (NDE) through a locked obstacle.

- Adjusting or customizing a tool during an entry, or using a commercial lock picking tool in a way that it is not intended, to help you gain NDE.

- Using a tool (entry tool or otherwise) for something other than its intended purpose.

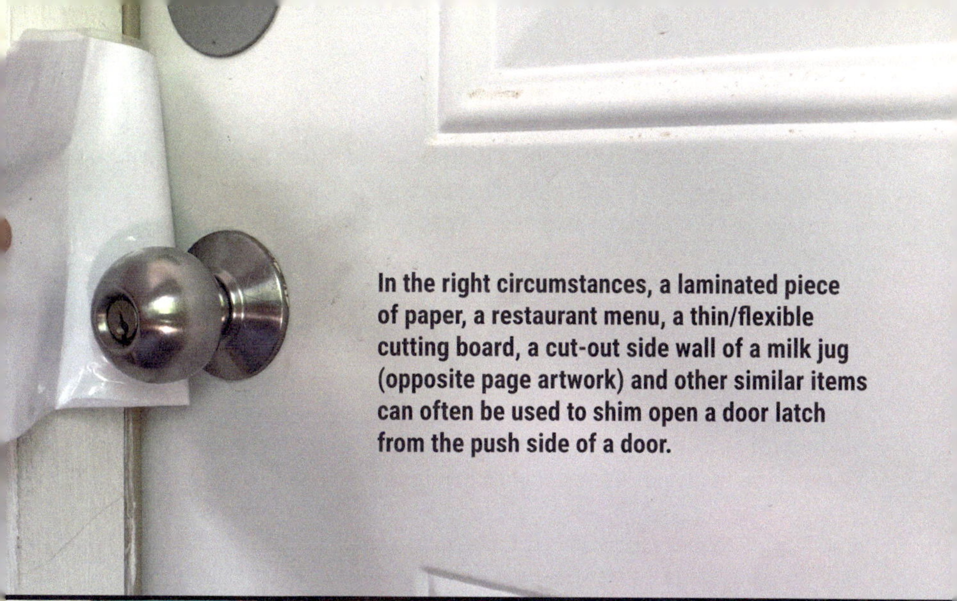

In the right circumstances, a laminated piece of paper, a restaurant menu, a thin/flexible cutting board, a cut-out side wall of a milk jug (opposite page artwork) and other similar items can often be used to shim open a door latch from the push side of a door.

Paint scraper. For slipping door latches from the pull side of a door.

Hair snap clips, barettes, that have the right size, shape, and texture, can often be used to open up the Master Lock model 175's and similar makes/models.

Utilizing makeshift tools in the field is not just about knowing a list of commonplace items that pair with a certain type of lock for an entry. Being able to leverage the concept of makeshift tools must start with a foundation of how the locking systems operate, inside and out. Understanding the basics of manufacturing and how to work with different materials in a *repeatable* and *reliable* method is also a necessity if you want to leverage makeshift tools into your toolbox for operations and emergencies.

The best thing you can do is try, and fail, and try again. Become familiar with which tools are the best at altering other materials. Become familiar with common methods for heating and melting things, common methods for cutting things, common methods for sanding and filing things.

And be sure to test your methods multiple times in a lab setting, before thinking that one success at home will equal a definite success in the field.

Try to remember not to get locked into a rigid thinking framework that "this company sells this tool, this tool is what I need in order to make an entry." There are many cases where literal garbage will get you a successful entry when a commercial entry tool won't.

Just because a specific tool is known to work on something doesn't mean it will work. Maybe a slight alteration would make that tool work better. Maybe a tool from another commercial purpose might actually work better.

Try to keep an open mind, and remember "What else could I use here?"

(Below: a thin metal #7 address marker, being used to slip a door latch)

A student using a blood pressure cuff as an air wedge.

THANK YOU

First and foremost, UTAC was born with a passion for teaching. We are always incredibly thankful when we can get our educational books into people's hands; and bonus points for *physical* books onto your physical bookshelves.

Remember that these are guidelines in this book, not rules. If something here isn't useful for you, or if you don't like it, throw it out. For the stuff you do like, take it and use it as you see fit. Make adjustments to your craft along the way.

Take caution when you learn about new tools/techniques and locks, as well as even the older tools/techniques/locks. You should be the one determining the value and the validity of every tool and technique that crosses your path. Your brain is the decision maker for how you build and how you apply your craft.

We are thankful for the entire community of lock pickers around the internet and around the world. We do like to innovate where we're able to but we also want to acknowledge that innovation often comes from being inspired by others. We pull a lot of our own data, we add a lot of our own context, and we often bring our innovations, small and large, to our students because we are inspired by the wonderful work of the entire community of entry studies.

Hobby lock pickers; locksport enthusiasts; firefighters, patrol officers, SWAT and special teams operators, red team and pentesting professionals, locksmiths, preparedness and survival minded individuals, homesteaders and DIY-ers, special forces operators; and anyone else that we missed, too; we are thankful that people out there care enough about the craft of entry studies to share their crafts with the world.

Please share our books and our methods with those who you think may benefit from it.

And we love having real humans come hang out with us at our training compound so please keep us in mind when you're planning your training events for the year. UTAC is currently hosting classes out of our private training facility located in MIDDLE Georgia. Students in our multi-day courses can sleep on-site in our bunk room for no extra charge. We are rural and far away from the hustle and bustle of big cities. And we are the only ones on site that have exclusive access to this facility. We are happy to offer a world-class level of instruction at a unique and inspiring training site.

Thank you, again.
-Pat & Team
Email: Pat@utac.io

Made in the USA
Monee, IL
12 September 2026

63115047R00067